VACATIONSCAPE
Developing Tourist Areas

Third Edition

Clare A. Gunn, Ph.D., F.A.S.L.A.

Professor Emeritus
Department of Recreation, Park
and Tourism Sciences
Texas A&M University

Taylor & Francis
Publishers since 1798

USA	Publishing Office:	Taylor & Francis 1101 Vermont Avenue, N.W., Suite 200 Washington, D.C. 20005-3521 Tel: (202) 289-2174 Fax: (202) 289-3665
	Distribution Center:	Taylor & Francis 1900 Frost Road, Suite 101 Bristol, PA 19007-1598 Tel: (215) 785-5800 Fax: (215) 785-5515
UK		Taylor & Francis Ltd. 1 Gunpowder Square London EC4A 3DE Tel: 171 583 0490 Fax: 171 583 0581

VACATIONSCAPE: Developing Tourist Areas, Third Edition

1 2 3 4 5 6 7 8 9 0   B R B R   9 8 7

The editors were Caroline Schweiter, Holly Seltzer, and Catherine Simon. Cover design by Michelle Fleitz. Prepress supervisor was Miriam Gonzalez.

A CIP catalog record for this book is available from the British Library.
⊝The paper in this publication meets the requirements of the ANSI Standard Z39.48-1984 (Permanence of Paper)

Library of Congress Cataloging-in-Publication Data

Gunn, Clare A.
    Vacationscape: developing tourist areas/Clare A. Gunn. — 3rd ed.
        p.   cm.
    Includes bibliographical references.
    1. Tourist trade.   I. Title.
G155.A1G863   1997
333.7—dc21   96-44472
                                                                     CIP

ISBN 1-56032-519-4 (case)
ISBN 1-56032-520-8 (paper)

Contents

Foreword

This third edition of *Vacationscape* remains true to the fundamental principles that made the first edition in 1972 an award-winning classic and yet includes current changes in tourism planning and development. The result is a volume of value to developers, planners, designers, politicians, students, and citizens alike.

When I first started focusing my professional career on tourism some twenty years ago, one of the pioneers whom I greatly admired and to whom I looked as a source of inspiration and knowledge was Dr. Gunn. His prolific writing and presentations often challenged the conventional wisdom of the time.

Over my professional career, I have had the privilege to know Dr. Gunn in a wide range of settings. I still remember most vividly the excitement of having him spend an afternoon in my home during which I simply listened, almost in awe, to the depth of his concern and understanding. This was at a time when tourism was still struggling to gain acceptance and recognition as a legitimate field of study. Whatever legitimacy it had achieved was in large part because of the writings of my guest.

At an age when most people have lost interest in their careers, perhaps even life itself, Dr. Gunn, as he enters his ninth decade, continues to seek, to explore, and to challenge many of the tenets that have been popular over the last fifty years. The vigor and continuing insight that he brings to tourism incite inspiration and indeed hope to many of us in the field who have relied on his writings to inform us, to guide us, and to challenge us.

In this edition, Dr. Gunn continues to meld the fundamentals of landscape architecture and planning with the field of tourism. His work continues to innovate in terms of both its orientation and its attention to detail. The result is a text that readers can be sure remains fundamentally sound while incorporating the societal, economic, ethical, and environmental changes that have taken place since the preivous edition in 1988. Despite his commitment to the basics, Dr. Gunn has retained an amazing capacity to understand our evolving environment and to interpret the significance of this evolution in a compact and lucid volume.

In addition to the basics that he has drawn from worldwide experience, this text includes current topics such as environmental impact, sustainability, and ecotourism. Special emphasis is given to the growing need for business to implement environmental protection and ecological integrity as an essential part of economic development, especially for the nature segment of the tourism market. His wisdom accepts the reality of the "in" concept of ecotourism at the same time he expresses concern over excessive proliferation of development in rare and fragile environmental settings.

Communities seeking tourism development will find here many crisp and clear guidelines for better success and avoidance of its many pitfalls. Especially important is the book's strong emphasis on community involvement and greater public–private cooperation, practices that are so often neglected in so many areas.

Another basic characteristic of *Vacationscape* is the manner in which it brings together research and practical applications of that research. Dr. Gunn's many years of experience have provided him with a clear understanding that research is essential for effective planning and development. His own research and principles derive not only from other scholars but from his own critical observation of the real world of tourism. His former students now apply his theories and principles in many parts of the world because they are realistic.

Those seeking justification for research without utility will not find solace in the present text. It is fundamentally a work that seeks to provide a strong bridge between theory and practice. This factor, combined with the scope and clarity of the book's presentation, makes *Vacationscape* a reference text of great value to students and practitioners as they seek to learn more about tourism. It is of value to all who need to know how best to resolve the practical issues of the day. This work fills a special need in today's literature on tourism.

—Dr. J. R. Brent Ritchie

Preface

Why *Vacationscape*? I coined the word during my visiting professorship in Hawaii in 1966. It stemmed from my basic profession, landscape architecture, and the escalating pastime of the day, vacationing—the American term for holiday or pleasure travel. Perhaps writings such as Gordon Cullen's *Townscape* in 1961 also had subliminal influence.

Because of my accidental entry into the field of tourism education in 1945, I have had the rare experience of witnessing five decades of tourism change. Over this period my sensitivity to natural beauty and man-made ugliness has not faded but rather has sharpened, almost to the extent of outrage. But complaining doesn't solve problems. Therefore my entire career has been directed toward learning, teaching, and writing about the vision of a more beautiful, a more satisfying, and a more adaptive landscape of travel.

Tourism is more than hotels, important as they are for traveler accommodation and economic impact. Tourism is more than roads, airlines, and ships, significant as they are for access to destinations. Tourism is more than promotion, in spite of the billions of dollars spent on travel advertising by business and government. All tourism takes place on land; and yet how little attention is paid to how that land is protected, planned, developed, and managed for tourism—that land through which all visitors flow. What do we as travelers see, smell, feel, and hear as we travel, and are designers and developers sufficiently sensitive to our interests and reactions?

Fortunately, today the answer to the latter question is increasingly a "yes"; the tourism–land interface is gaining much more attention. Whereas it took two years to find a publisher for the iconoclastic first edition of *Vacationscape*, today's publishers eagerly seek writings on tourism and the environment. Over the years the dramatic growth of research, education, and scholarly writing about tourism, once considered soft and frivolous, has been astounding.

Why a third edition? Since my retirement from the rigors of the class schedules, examinations, and faculty meetings demanded by academe, release time has afforded me more global travel, more reflection, and more consulting. In the years since the first and second editions, during which I wrote a comprehensive third edition of *Tourism Planning* (1994), the need became evident for a book more clearly directed toward environmentally sensitive development, especially at the local level. In spite of several worthwhile "community tourism" guides, the linking of the role of designers and developers to tourism was not being fulfilled. Hence this third edition of *Vacationscape*.

But readers will not find complete solutions to all the ills of bad tourism development here, because no one has them. Instead, it is my hope that with this writing, all players can be stimulated to hone their own skills, intuition, talent, and good judgment to strive toward better solutions. Tourism is with us and will not go away. It continues to grow and absorb many more sections of pristine land every year. In so doing, it demands far better awareness and creativity than it is getting today. There must be better ways of protecting environmental assets as more people enjoy them; providing greater connectivity among all development; ensuring less culture shock between hosts and guests; enhancing, not detracting from, local ways of life; and guaranteeing future populations the last vestige of freedom and adventure remaining—the rich, invigorating, pleasurable, and enlightening experience of travel.

As in the first edition (1972), readers will find here much *description*, especially of tourism development fundamentals that are universal and as fresh and applicable today as they were more than two decades ago. In spite of differences in traditions, governance, and degree of economic development among nations, the same basic elements of tourism—transportation, attractions, and services, all built on land—must be dealt with everywhere.

But without some *prescription*—concepts, ideas, visions, ethics—applied to these fundamentals, there can be no improvement. Therefore, I make no apology for my equal emphasis on advocacy based on observation and experience. Such recommendations are not capricious conceptions. They are conclusions, meant to be constructive for all—students, designers, developers, administrators, and especially travelers. Not all have been tried; they just seem to be natural outgrowths of the strong desire to create a better tourism world.

For this project I have had a great amount of help. Fortunately, today more literature on tourism is available, and I am indebted to many who have provided recent research, information, and case descriptions. I especially want to thank the following: Ann Checkley, Canadian Pacific Hotels Corporation; Kathleen L. Andereck, Arizona State University West; John Ap, Hong Kong Polytechnic University; Graham Brown, University of New England, New South Wales; David Botterill, Cardiff Institute of Higher Learning; Brian Hay, Scottish Tourist Board; Safei El-Deen Hamed, World Bank; Valene Smith, California State University, Chico; Donald Getz, University of Calgary; Stephen Hill and Pamela Wight, Alberta Economic Development and Tourism; Glenn Carls, consultant; Chaiyasit Dankittikul, Silpakorn University, Bangkok; Michael Fagence, University of Queensland; Ann Keng Montgomery, Texas A&M University; Keith Hollinshead, Texas A&M University; Peter Hawley, National Endowment for the Arts; Robert McNulty, Partners for Livable Communities; Richard Gitelson, Pennsylvania State University; Scott Meis, Canadian Tourism Commission; Ronald W. Tuttle, Natural Resources Conservation Service; Kenneth Hornback, National Park Service; Paul Leeson, Purcell Lodge, British Columbia; James H. Wies, Shearwater, Nova Scotia; Donald F. Hilderbrandt and John C. Hall, LDR International; Dixi Carrillo, EDAW, Inc.; Noel Viljoen, Chris Mulder Associates, Inc., South Africa; Pieter Bekker, Ministry of Small Business, Tourism and Culture, British Columbia; and Carla Short, Conservation International.

I much appreciate the information and case studies supplied for the second edition that were adapted for this volume. The photographic copy work by my son Richard of Commercial Images is also appreciated. And, finally, this book would not have been possible without the constructive support and many hours of word processing by my wife, Mary Alice.

—Clare A. Gunn

Tourism: Positive and Negative

A most difficult lesson of tourism development today concerns the recognition that tourism has both dark and bright sides. Today the greatest wave of land development worldwide is aimed at tourism. As travel demand increases, nations, provinces, and communities look to tourism for economic progress, even salvation. It is reported to be the largest component of world trade, employing over 212 million people, one in nine workers. It now represents more than 10 percent of all global wages and is expected to continue its expansive growth (World Travel & Tourism Council 1995).

When tourism is not planned and developed so that it avoids, or at least reduces, its negative impact, it does exact some costs. Following is a sketch of significant issues that can become pitfalls if not considered at the very beginning of tourism development. Most of the difficulties are attributable to a lack of understanding and planning at the outset. Historical review suggests that poor preparedness is the greatest cause of negative impact.

HOST–GUEST RELATIONSHIPS

Today, neither the distribution nor the degree of tourism development occurs equally around the world. Some destinations have reached saturation, whereas others have not been visited by anyone. Even those areas with extensive physical development for masses of tourists seldom have well documented histories of why or how their tourist development occurred. Seldom do these destinations have sophisticated organizations or policies to manage tourism. Even where concerted planning and management have taken place, technological transfer to other regions has been weak and erratic. In spite of the many studies and treatises, mostly on promotion, the question of how to develop tourism seems to remain a mystery. Areas newly exposed to visitors frequently are ill-prepared to receive them, especially when they come in massive numbers. One can speculate on the many reasons for this great gap between hosts and guests, but experience suggests a few major factors.

Perhaps foremost is the lack of understanding of tourism by residents at the local development level. It is often thought of as the business of someone else or merely ignored completely. Past economic development usually has not provided the training or experience to give the local people any insight into the realm of tourism. Most development in the world grew from ancient food gathering and hunting into other economic foundations such as agriculture, mining, forestry, and fishing (Bronowski 1973). Technical and practical information about such activities was well known locally and dominated all social and economic life. Other than receiving guests in their homes, local people had no experience with the concept of tourism. The potentially great complexity and deep penetration of the many tourism tentacles into a community were unknown.

Tourism development grows from external intrusions rather than from local generations. Geographer Pearce (1989) has cited Miossec's model (1977) as perhaps the most lucid geographical description of tourism development (Figure 1-1). It reveals a progression from tourism's infancy, with just a resort or two, through maturity of greater saturation. It also identifies changes in transportation, tourist behavior, and the attitudes of decision makers and local populations. Pearce applied this model to Mediterranean tourism and found that the sequence of stages held for that area. This model emphasizes the dynamics of tourism, the changes in both hosts and guests, and the consequent impacts at each stage. Others, such as Butler (1980), have modeled tourism development in different stages: exploration, involvement, development, consolidation, stagnation, and rejuvenation or decline.

Although these models explain tourism's evolution, they are not useful for local tourism planning. Typically, development came from foreign investors who saw opportunities, purchased land, and proceeded with development without input from local residents. Only a few

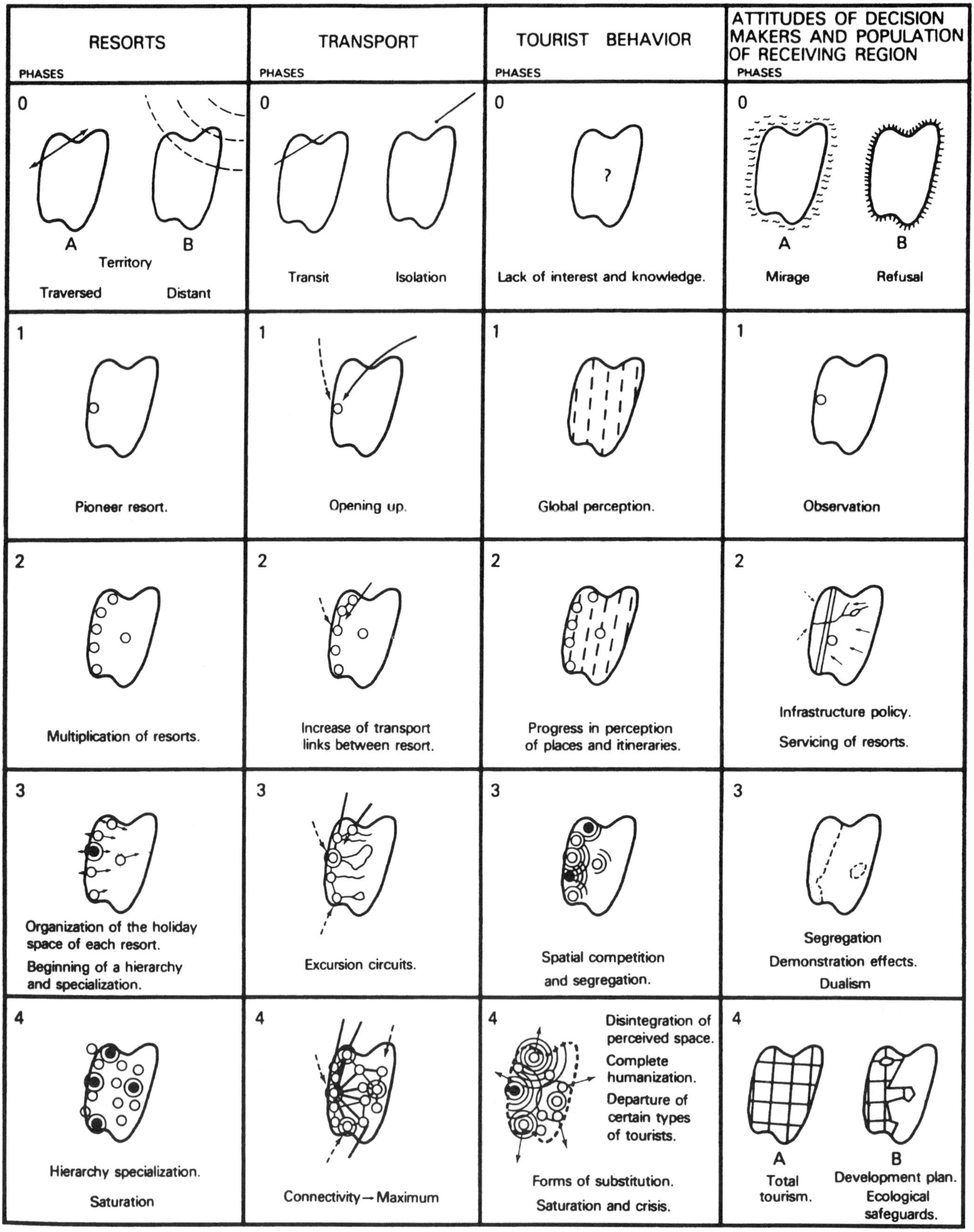

RESORTS	TRANSPORT	TOURIST BEHAVIOR	ATTITUDES OF DECISION MAKERS AND POPULATION OF RECEIVING REGION
PHASES	PHASES	PHASES	PHASES
0 A B Territory Traversed Distant	**0** Transit Isolation	**0** Lack of interest and knowledge.	**0** A B Mirage Refusal
1 Pioneer resort.	**1** Opening up.	**1** Global perception.	**1** Observation
2 Multiplication of resorts.	**2** Increase of transport links between resort.	**2** Progress in perception of places and itineraries.	**2** Infrastructure policy. Servicing of resorts.
3 Organization of the holiday space of each resort. Beginning of a hierarchy and specialization.	**3** Excursion circuits.	**3** Spatial competition and segregation.	**3** Segregation Demonstration effects. Dualism
4 Hierarchy specialization. Saturation	**4** Connectivity—Maximum	**4** Disintegration of perceived space. Complete humanization. Departure of certain types of tourists. Forms of substitution. Saturation and crisis.	**4** A B Total Development plan. tourism. Ecological safeguards.

Figure 1-1. *Growth of tourism development. A model of typical changes with growth of resorts, transportation, tourist behavior, and host reception (Pearce 1989, p. 17).*

local landowners and public officials were typically in on the deal. Occasionally, a few local people saw opportunities for providing services to visitors and established facilities. Generally no comprehensive involvement took place to give everyone understanding of the advantages and potential disadvantages of tourism development. Nor were the locals given any say in decisions. As a result, they often faced environmental degradation, traffic congestion, economic and social disruption, and other ills from mass unplanned tourism development. Unfortunately, this pattern of evolvement is standard today.

Because tourism development occurred somewhat differently in cities than in rural areas, small towns, and primitive areas, it is useful to examine the sequence of development for each. From this examination of the past, tourism planners may learn to avoid development pitfalls and to visualize new and better approaches.

Cities

Cities throughout the world were created primarily as physical centers of life where work, play, trade, and social exchange could take place more effectively than in the hinterland. As such, they also became centers of culture, education, and worship. Physical development of land for cities was an expression not only of needed functions but also of the desire for beauty, repose, and pleasure.

For self-interest, citizens created a great many amenities that became a fundamental part of urban fabric, such as:

> architecture, building patterns, landscape features;
> public spaces, plazas, streetscapes, gathering places;
> opportunities for people-watching;
> places of street entertainment;
> parks, gardens, recreation areas;
> zoos, aquariums, museums;
> theaters, places for entertainment, storytelling venues;
> farmers' markets;
> shopping areas;
> sidewalk cafes, street vendors;
> cultural and historic sites, commemorative places;
> festivals, craft displays, ethnic customs; and
> sculpture, art.

These places were often given further support from several sources. In many nations, park, recreation, literary, scientific, and historic societies rallied behind the protection and management of these amenities as early as the eighteenth and nineteenth centuries (Ashworth and Tunbridge 1990). This support occurred especially in urban centers such as London, Berlin, and Paris. Later, local and national legislative measures gave further protection and advancement to these amenities, primarily as a citizen good.

Other features became essential parts of cities. All transportation termini (air, land, water) were established at cities. Internal transportation was provided by vehicles and systems—first horse-drawn coaches and then taxis, personal cars, buses, mass transit, and rental cars. Food, automobile, and travel services were built in cities. Health and safety services evolved there—doctors, dentists, emergency care, hospitals. Public services such as police, public water and waste systems, and fire protection were provided in cities. Housing, schools, universities, churches, and auditoriums were founded there. All these physical developments were created for the use of local residents and supported by their tax monies.

Local pride and loyalty to cities and small towns must not be overlooked as a social fundamental throughout history. Essential to citizens of cities in many parts of the world today are the elements that provide opportunities for assembly, interaction, and even emotional attachment to place, especially in areas of more stable populations. In a mobile country such as the United States, this fervent support and even defense of local place is less evident. In western Europe, on the other hand, following the devastation of many cities in World War II, the remaining population did not flee from their homes but immediately began rebuilding (Lennard and Lennard 1995). The very first effort was to restore the meaning of community life. As everywhere, when threats to the fundamental character of cities occurred, the potential loss to human stability and continuity became real.

Then came tourism, an onslaught of outsiders. Travel marketers, both inside and outside the cities, began to extol their urban features. Hordes of visitors came, often a thousand times the local population. Generally, cities were ill-prepared for this invasion. Those originally responsible for the development and management of the amenities had no part in inviting so many of these visitors to their sites. Those who fanned the flames of advertising and other forms of promotion had no part in the decision-making process within destinations. The local citizens who were affected by masses of tourists likewise had no voice in decision making. Thus appeared a great gap between those who induced travel and those responsible for managing amenities in destination cities.

Rural Areas and Small Towns

The gap between those promoting travel and those who lived in and managed lands has often been greater in rural areas and small towns than in cities. These lands were often under control of government agencies, private firms, and local citizens who were involved not with tourism but with mining, forestry, agriculture, or extensive parkland and preserve activities. The interests of the small town administrations and citizenry were focused predominantly on the activities surrounding these uses of the land. Often included within their extensive lands were rivers, lakes, waterfalls, valleys, hills, mountains, deserts, and wildlife habitats, and sometimes aboriginal or prehistoric sites. For rural and small town dwellers, these amenities were traditionally *theirs*.

Then came volumes of visitors seeking the recreational and cultural values these resources could offer. Hundreds of thousands, even millions of visitors began to come each year to participate in activities such as the following: camping, nature study; hunting, fishing; hiking, walking, backpacking; mountain climbing, mountain biking, skiing; rafting, boating, swimming; photography; resorting, vacation home use, condo use; ghost town visits, small town historic visits; and dude ranching.

Even less prepared than cities were these resource areas, their leaders, and residents. Seldom did they have in place the needed zoning statutes, plans, planning officials, or controls to avoid catastrophic results. The hordes of visitor populations from cities generally held personal lifestyles and interests quite different from those of rural residents. Visitors invaded lands and resources that for generations had been considered solely the turf of locals. Often, the administrative offices that governed these extensive lands were located long distances away, making access and political influence by locals extremely difficult. Certainly the local vocational interests and traditions were much in contrast to those of the visitors. As a consequence, many local resources in rural areas were either destroyed or badly damaged.

Nonindustrialized Areas

In undeveloped regions of the world, many people still live on subsistence agriculture and by barter rather than in a money economy. In many areas they occupy lands abundant in fish, wildlife, and forests. Land and natural resources are essential parts of their lives. Over many generations, many customs, crafts, and worship rituals have become essential parts of their cultures.

Because of the abundance of exotic animals, rare cultural customs, and often spectacular scenic beauty, in recent decades tourists have begun to invade such areas. Despite language barriers and primitive food and sanitation standards, tourists have begun to trample the environment, slaughter the wildlife excessively for trophies, and impose on the privacy of native peoples.

Again, the management and control of resources by local or national governments have often been lacking. Safari tours in Africa, for example, became so intensive an intrusion that the very wildlife that was the area's major attraction was severely threatened.

CONSEQUENCES

In all these situations (cities, rural areas and small towns, undeveloped regions) there is ample evidence of tourism's erosiveness. Further examination shows that such damage is preventable—but not without new understandings, planning policies, and actions. As it continues to spiral upward, tourism continues to consume vast acreages in ways that make the land very difficult to reclaim. New resorts, new airports, new roads, and new shops, casinos, and hotels demand the very best of available land resources. Although tourism may not make as drastic nor as permanent an impact on land as do mining and other extractive industries, its effects are real and worthy of consideration everywhere that continued growth is contemplated. Equally vulnerable are all categories of places—cities, rural areas and small towns, and developing nations.

Physical Resource Degradation

Increasingly, the negative impacts of tourism on the physical environment are being recognized and dealt with in a constructive manner. Wight (1994) has cited many instances in which commercial lodging and food service businesses are becoming more environmentally friendly. Although a moral obligation may provide one motive, probably much stronger motives are the demands by consumers and cost savings to developers. For example, by implementing conservation measures, Italy's Hotel Ariston increased its occupancy by 15 percent, and the Boston Park Plaza Hotel increased its business by $750,000 a year. The Royal Connaught Hotel in Hamilton, Ontario, saved $33,113 annually by replacing conventional lights with compact fluorescents. Prince Edward Hotel in Charlottetown, Prince Edward Island, reduced fuel consumption by 97,000 liters in one year by installing a water source heat pump (Wight 1994). Generally, the conservation measures being implemented by tourist service businesses can be classified into recycled and reduced waste, conservation of energy and water, improved air quality, or more selective purchases of products. There are many other instances of tourism development that are actually enhancing local environments by diverting some business profits toward support of nearby park and natural resource areas, such as ecotours.

But, at the same time, natural resources, the very foundations for tourism, are being eroded as never before. Especially damaging is water pollution, because clean water is essential to so many tourist activities. A few examples illustrate the seriousness of water contamination, not only by tourism but by industrial and municipal waste: More than 2.5 billion gallons of untreated waste is flushed annually into Narragansett Bay, Rhode Island (Brancatelli 1995). In 1995, 9,942 volunteers for the Center for Marine Conservation collected 361,000 pounds of debris along 147 miles of the Texas Gulf Coast (1995 International Coastal Cleanup, 1996, 1). This is not merely an American problem; it is worldwide.

Resort hotel sewage often flows untreated into the recreational waters at its doors, threatening the health of all tourists who swim and fish there.

Some observers are concerned about the ability of tourism to completely change the character of cities. Selzano (1991) has cited the case of tourists to Venice, where more than six million visitors a year descend upon a city of only eighty thousand inhabitants. As a result, tourism activities have pushed out local businesses such as shoemakers, plumbers, bakers, butchers, milliners, and haberdashers that formerly served the residents; tourist-type shops have replaced them. The mayor, Antonio Casellati (1991), is of the opinion that numbers of tourists must be restricted, especially on a site-to-site basis. He turned down a proposal for a World Fair (Expo 2000) that would have increased the volume of visitors by two hundred thousand per day.

Social and Cultural Resource Erosion

For generations, traveling has been accepted as a culturally broadening experience. The "grand tour" of the fifteenth century throughout southern Europe was recognized by British aristocracy as a necessary part of a gentleman's education (Hibbert 1969). Ever since, the culturally enriching aspects of travel to other nations have been well recognized. Even during the Cold War between the United States and the Soviet Union, President Reagan stated, "People-to-people contacts can build genuine constituencies for peace in both countries" (D'Amore 1988, 12). D'Amore (1988) quoted Pope John Paul II as saying:

> The world is becoming a global village in which people from different continents are made to feel like next door neighbors. In facilitating more authentic social relationships between individuals, tourism can help overcome many real prejudices, and foster new bonds of fraternity. In this sense tourism has become a real force for world peace. (13)

Even though a great amount of tourist travel has been trivialized and is superficial, the benefits of meeting and observing other cultures remains a positive force within tourism.

However, native structures, land uses, customs, and private values are extremely vulnerable to invasions by those of contrasting cultural backgrounds. Contrasting lifestyles practices, such as nude bathing, sports dress in restaurants, and tourist desires for massage parlors, drug sales, strip bars, and prostitution, have dramatically upset local cultural traditions in many destinations. More insidious has been the destruction of historic structures, prehistoric artifacts, and indigenous forms of entertainment and crafts by well-meaning but thoughtless tourism developers. Even festivals, often touted for their cultural exchange opportunities, often fail to meet these expectations (Roche 1994). Anthropologist Smith (1977), in her classic volume on tourism's host–guest relationships worldwide, concluded that in the great drive toward expanded tourism economics, a significant amount of negative cultural impact is taking place, but that this is not irremedial. For example, in its zeal to develop new tourism, The People's Republic of China has created a new theme park in Beijing called "World Park" that contains replicas of the Sydney Opera House, the Statue of Liberty, the Grand Canyon, Great Pyramid in Egypt, and the Taj Mahal. Sun Xiao Xiang (1995), a tourism planner and scholar, has expressed concern over the emphasis on building such imitations instead of expending talent and policy toward protection and public use of China's own rare natural and cultural features.

Especially vulnerable to tourism's cultural impact have been the undeveloped and developing countries of the world. In places such as Gambia, Thailand, and other Asian destinations, prostitution has become the dominant image of the travel purpose (Lea 1988). Young women and boys from rural areas are frequently sold to the brothels of the larger cities. Governments are now making strong efforts to resist these consequences of tourism but may find economic and market demand pressures hard to overcome.

Hollinshead (1996) expresses great concern over tourism exploitation of Aboriginal culture in Australia, which is a result of the recent surge of access to this indigenous population. Marketing this culture tends to be insensitive to the rich metaphysical traditions of this unusual society. Hollinshead advocates protected anthropological zones in which cultural ecosystems of these special populations are designated and properly interpreted, if tourism penetration is allowed.

Economic Costs

Throughout the world, the greatest motivation for promoting tourism is its economic value. Modern research has demonstrated its value in creating jobs, providing income, increasing tax revenues, and generating investment. In the state of Texas, for example, it was reported that in 1993 tourism added $22.9 billion into the state's economy and employed approximately 422,000 individuals (Texas Tourism Division 1995). Even though domestic (U.S.) travelers made the greatest impact, Texas was a leading destination for international travelers—4.7 million in 1993, 10 percent of all international travelers to the United States. On a smaller scale, the community of Sandpoint, Idaho (population 6,500) has reduced unemployment, increased per capita income, and stimulated other industry by increasing its tourism activity by 115% (Koth et al., 1991).

Of course, tourism development is not free of costs. Experience has demonstrated several categories of economic costs. With tourism expansion come new demands on transportation systems to handle large volumes of traffic. The establishment of Walt Disney World, although it is now internally self-sufficient, required a public investment of $5,000,000 in new access highways (Weaver 1991). For new tourism growth, other types of transportation investment may be required, such as new buses, taxis, and rental car inventories. Local streets and rural roads may need to be widened, resurfaced, or relocated to accommodate greater tourist volumes. Because of the increased trend toward automobile and motorcoach use by day tourists, a major demand may be placed on investment in new parking areas. For directional purposes, new signage may be required.

Another major new local demand is on public services such as water supply, waste disposal (sanitary and solid waste), police services, and fire protection. Usually small communities do not have excess capacity of these systems and require new taxation to expand these systems when tourism is developed.

For many years, nations and communities have predicated policies for tourism on growth—the more visitors the better. Now, this postulate is being questioned. Although many areas have demonstrated great elasticity, others have reached the point where the social, environmental, and economic costs are raising questions about their continued growth. For example, after analyzing New Zealand's tourism, Getz (1994, 3) observed, "It can be concluded that the industry is not better off when it constantly pursues growth, either on its own accord or on the behest of government and tourism development agencies." Instead, there are increasing pleas for quality, not only for avoiding local degradation from excessive erosion from masses of tourists, but especially for the quality of the visitor's experience. Too many tourists in one place often translate into inferior sightseeing, traffic gridlock, and exhaustion because of the difficulty of reaching target attractions. The trend toward ecotourism illustrates this risk—if too many tourists are in a rare natural area, both the visitor experience and the environment will lose. Instead, many now believe that increased quality rather than quantity has much merit, perhaps even greater economic yield.

Disunited Development

As thousands of decision makers continue to create new attractions, services, and transportation networks, the result is a chaotic mass of unconnected development. In their haste to develop, individual land developers have not integrated their actions with others', often creating a confusing and ugly juxtaposition of land uses. As geographer David Lowenthal (Lowenthal and Prince 1955, 82) has stated, "Features of interest often lack all connectivity." After many years of extensive travel in Egypt, landscape architect Hans Friedrich Werkmeister (1986, 182) reported on the "carelessness and even the

shamelessness by which so-called responsible people are pushing development of tourism to its most awkward lengths." The blessing of unlimited access compounds the problem by compressing large numbers of developments into the traveler's experience. This situation results in less than optimal business success, poor visitor satisfaction, and conflict with existing community values, as well as environmental degradation. Few community or national policies have been created to guide tourism growth in directions to provide a more integrated and more smoothly functioning whole. Enclave tourism, such as beach resort development, is entirely market-driven and created with no relationship to local culture, transportation networks, or other economic development, often creating a drain on the overall economy rather than enhancing it.

Homogenization

Mass tourism today continues on a path of sameness—a homogenized aesthetic landscape. Business developers everywhere tend to copy others who have appeared to be successful. Franchise and chain firms tend to repeat the same land use and building design in all locations and geographic regions. In the midst of this great desire to standardize offerings, the visual landscape of tourism is beginning to look all alike, hardly worth leaving home for. Although no one wishes to censor site development and building styling, design taste and judgment suggest the great need for greater sensitivity to geographic and cultural differences—the inherent causes for most travel.

Destruction of Place Qualities

Closely related to homogenized development is the destruction of place qualities. Hills are bulldozed flat, native vegetation is destroyed, and exotic plantings are added. Often vernacular architecture is replaced by international glass-box style, historic places are destroyed by gambling casinos, prehistoric sites are buried under golf courses, and significant vistas are blocked. Place qualities, the very essence of the pulling power of destinations, are unwittingly being decimated. After visiting the Nile, landscape architect Adolf Schmitt (1986, 158) observed:

As the objects worth seeing are slowly given way to destruction, the one-sided development of tourism leads itself to absurdum. First of all, exaggerated traffic handling: the "Valley of the Kings" is covered with asphalt and concrete right into the center, which today is occupied by a monstrous restaurant building, suitably called the "Tomb of Coca Cola." There the tumultuous eating and noisily drinking human mass is sitting right in the middle of one of the most time-honored and most ancient graveyards of this world.

These practices are destructive not only for sound and long-range tourism but also for maintaining the local quality of life. Until the damage is done, local citizens are generally unaware of how their homelands, important to them for generations, are being drastically altered, all in the name of tourism progress. Of great importance to place quality, especially for cities, is the historic background. So-called modernism and progress for tourism tend not only to remove important elements of the past but also to intrude upon the historic city that visitors came to see (Ashworth and Tunbridge 1990). Developers often forget that everything created for the city has two kinds of users—residents and visitors. To satisfy both requires considerable communicative exchange and often diplomatic compromise. Today's newly acquired Black leadership of South Africa, for example, may not view the Voortrekker Monument at Pretoria—a shrine commemorating the bloody battles and massacres that resulted in White supremacy in the Great Trek of 1838—as a desirable artifact to show tourists.

COMMUNITY CHOICES

Even though much pressure for tourism development comes from outside, local communities can exercise much greater power of choice than they may now recognize. Their choices may be placed into three broad categories: prohibition, capitulation, and adaptation.

Prohibition

Although difficult to accomplish, local people should have the right to say, "No, we don't want any tourism development." It can create a lot of litter and congestion of traffic, conflict between foreign visitors and local residents, increases in land values, and even disruption of local lifestyle. For example, in 1993, the community of Byron Bay, Australia, began to voice serious concerns about the proposed establishment of a new Club Med resort there (Heaney, 1993). Heaney stated, speaking for the local residents, that "we do have the right to make decisions for the benefit of the community and that we don't have to accept that the speculator and the multinational have a God-given right to profit at our expense" (Heaney, 1993, 1). A local organization called Byron Shire Businesses for the Future stated its objectives as:

1. to conserve and enhance the environment in Byron Shire,
2. to encourage the success of locally owned and operated businesses within a diversified and broad-based economy,
3. to foster a high quality of life-style for residents of the Byron Shire,
4. to gather and disseminate information on developments within the Shire, and

5. to encourage participation of the community in the development process. (Heaney 1993, 1)

The expressed concerns over the proposed resort included: less job creation than promised, encroachment upon local quality of life, tourism becoming a monoeconomy, destruction of the area's wildlife habitat, environmental degradation of the proposed site, and economic drain in favor of outsiders.

To prohibit tourism development requires severe control and local policy action. In most communities, prohibition has not been seen as a viable option because of its overwhelming reputation for positive impact on the local economy. But this decision generally has not been based on a thorough examination of the consequences of development, such as in the example of Byron.

Capitulation

In most instances around the world to date, local areas have given in to any tourism development by anyone, because it has been seen as promising positive economic impact such as new employment, new income, and new tax revenues. However, in hindsight, merely giving in to any development has yielded some less than desirable consequences, as already described.

Often coastal development, popular with guests, has been placed too close to the water's edge, resulting in storm and wave damage to beaches and hotels. In a great many instances, untreated sewage from resort hotels has been allowed to flow directly onto the beach area, resulting in pollution of the very waters used by guests for swimming, scuba diving, and other water sports. Lack of sensitivity to land resources has frequently resulted in erosion of soils, destruction of beautiful vistas, and depletion of wildlife habitat, all essential to many tourist activities. Tourism development has often been built upon the very natural and cultural resource areas that should have been protected as attractions.

The willy-nilly approach to tourism, so dominant thus far, too often has not produced the economic returns promised and too often has resulted in many negative impacts, not because of tourism per se but because it was not understood fully and planned to meet local as well as visitor needs.

Adaptation

A third choice, one that appears to produce the best results, is that of adaptation—constructively adapting tourism to the local situation. This approach requires a great amount of local self-discipline but is most satisfactory in the long run. It places decisions locally, so that communities and their surrounding areas can analyze their resources and goals, and make recommendations on where and how tourism should be developed, for their best interest as well as that of tourism. Virtually all of the negative impacts can be avoided when communities take the responsibility for guiding tourism growth in directions best suited to the local situation. Only when this is the approach taken will communities be able to respond to outside influences for tourism development, such as from government or outside investors. Their proposals may be desirable, but only when screened by a knowledgeable local populace. A striking example of a community that has taken control of its own tourism destiny is the island of Providencia, located in the Caribbean off the coast of Nicaragua. Prompted by a local civic association, the Ministry of Economic Development has declared a policy of low-key and locally controlled tourism development that protects the natural and cultural assets of the island (Schapiro 1995). It is toward such desirable ends that the many recommendations provided in this volume are directed.

SUSTAINABLE TOURISM

In recent years, the concept of sustainability has been put forward as an ideal that avoids the ills and also fosters the better goals of tourism development. This concept goes beyond the earlier philosophy of conservation, reaching into social and economic as well as resource dimensions. Following is a definition that can be applied to tourism with predictable positive consequences (Rees 1989, 13):

> Sustainable development is positive socioeconomic change that does not undermine the ecological and social systems upon which communities and societies are dependent. Its successful implementation requires integrated policy, planning, and social learning processes; its political viability depends on the full support of the people it affects through their governments, their social institutions, and their private activities.

When this definition is applied to tourism, it can be seen readily that relying only on market information is of minimum help. Instead, the emphasis must be on the developmental side, on being accountable for what is done and how it is done. Ritchie (1991, 97) has stated, "In the end, however, the political process and the power of different political units will determine the level and form which sustainability will take. Those of us in the tourism sector have traditionally ignored this reality and we are the weaker for it." Canada has been a leader in striving toward tourism sustainability, as shown in its *Globe '90* paper: "Sustainable tourism development can be thought of as meeting the needs of present tourists and host regions while protecting and enhancing opportunity for the future" (Tourism Stream Action Strategy Committee 1990).

This definition was interpreted by Taylor (1991, 28) to require action depending on a site's location along three dimensions:

1. The types of tourism: all will place different pressures upon the resource base;
2. The types of resource bases: each will be able to sustain different levels of development depending upon the type of tourism selected for a particular site; and
3. A ratio between actual visitation to a site and a level of sustainable visitation.

Taylor concluded that such guidelines demand that success "not be measured in more and more visitors, but in how well the level of sustainable visitation is maintained." At this same time, Ritchie (1991) outlined responsibilities of several organizations and levels in efforts to strive toward sustainable tourism development (Table 1-1).

Of great assistance to local areas is the acceptance of new environmental responsibility by businesses and large firms related to tourism. The World Travel & Tourism

TABLE 1-1. Responsibilities for Sustainability

Level/Organization	Responsibilities
Host community/ Region	Defining the tourism philosophy and vision for the community/region
	Establishing social, physical, and cultural carrying capacity for the host community/region
Destination management organization	Coordination of implementation of community sustainable development plan for tourism
	Monitoring of levels and impact of tourism in the community/region
Individual tourism firms and operators	Fair contribution to implementation of sustainable development plan for tourism
	Observance of regulations, guidelines and practices for sustainable development plan
Host community/ Region residents	Encouragement/acceptance of tourism within parameters of sustainable development
Visitors/Tourists	Acceptance of responsibility for minimal self-education with respect to values of host region
	Acceptance and observance of terms and conditions of host community sustainable development plan for tourism

Source: Ritchie 1991, 99.

This design demonstrates skill in balancing heavy tourist use with sustainable development. (Photo of Spanish River Park, Boca Raton, Florida, courtesy Edward D. Stone, Jr., and Associates)

Council (WTTC) has established a World Travel & Tourism Environmental Research Centre (WTTERC; WTTC 1993). It has set as its goals to "improve the environmental efficiency of existing operations, ameliorate past mistakes, and ensure that new developments are environmentally compatible." Its report cites several initiatives already in place by several major tourism firms. Thirty-five major global companies that account for travel, entertainment, and accommodation of several hundred million travelers a year have established new environmental policies. Among the key concerns of their mission statements are the following: waste management, energy consumption, staff training, community impact, emissions, purchasing policy, regulatory compliance, communication, hazardous materials, sustainable development, self-regulation, and water conservation/management. Canadian Pacific Hotels and Resorts were among the first to develop internal management practices more environmentally sound and sensitive to local community

interests (Troyer 1992). Following internal evaluations from ten thousand employees, a program was developed for the company's management staffs, in which the employees expressed their desire to:

- get in sync with nature;
- use our skills and knowledge to build a safe, clean-and-green future for our children; and
- become industry leaders—and planet savers. (Troyer 1992, 1)

The WTTC has cited three approaches for mitigating the local impacts of tourist invasions. International and national regulation may be necessary in some instances, such as for emissions, taxes, and effluent treatment. Local planning and control can guide the use of land, reduce environmental damage, and protect consumer interests. Business self-regulation is being implemented more extensively because businesses are now recognizing it is key to development. On the basis of a WTTERC study, Hawkins and Middleton (1993, 169) report the following themes among 16 major tourism companies that have adopted environmental policies:

- 75% state a commitment to reducing energy consumption and preserving non-renewable fossil fuels;
- 69% state that they will accept corporate responsibility for environmental issues and 38% aim to actively include such issues within corporate strategy;
- 69% state an aim to rationalize waste management systems and minimize waste generation (usually by the principle of the three "Rs": reduce, reuse, recycle);
- 44% aim to minimize or eliminate hazardous materials (including PCBs, CFCs, and pesticides) and a number of companies have taken considerable steps to develop alternative techniques;
- 38% aim to rationalize their use of water;
- 31% aim to maintain good environmental conditions to promote health and safety of employees and/or the public;

- 13% state an aim to involve the community in the operation and development decisions; and
- only one company explicitly stated a formal commitment to sustainable development in the formal mission statements available in mid-1992 (although statements about use of non-renewable resources and corporate responsibility have an obvious bearing on sustainable issues).

A major issue of sustainability centers on developing tourism and yet retaining elements of past local culture. In areas of the world where dramatic acculturation is taking place and older customs are being modified or abandoned, the question of offering visitors insight into these older cultural images becomes a development issue. Hollinshead (1993, 656) has argued that local areas have the "indigenous right for contemporaneity." Local residents have the right to progress. But to attract tourists, who often seek out destinations because of their exotic ethnic customs of the past, there may be no other solution than creating an attraction that preserves arti-

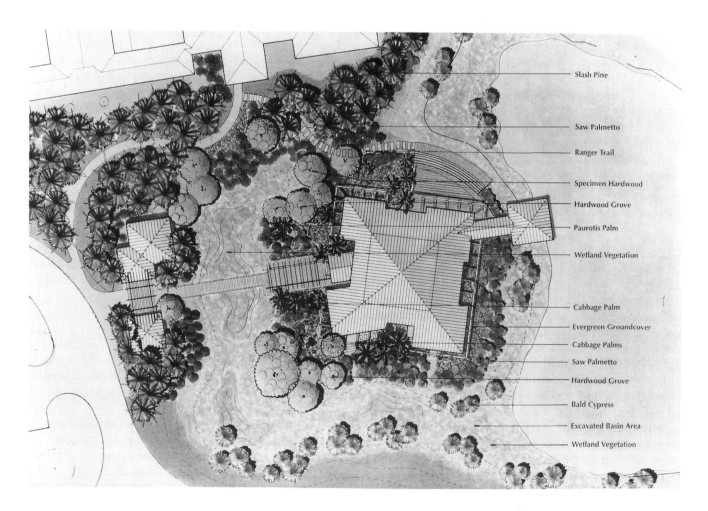

Slash Pine

Saw Palmetto

Ranger Trail

Specimen Hardwood

Hardwood Grove

Paurotis Palm

Wetland Vegetation

Cabbage Palm

Evergreen Groundcover

Cabbage Palms

Saw Palmetto

Hardwood Grove

Bald Cypress

Excavated Basin Area

Wetland Vegetation

Plan for new visitor center to replace inferior structure and site development. This sustainable design is for Everglades National Park. (Plan courtesy LDR International)

Computer-generated rendering of new visitor center at Everglades National Park, showing relationship to site. (Plan courtesy LDR International)

facts of that past. The Polynesian Culture Center of Hawaii, an attraction that depicts six historic Pacific cultures, is an example of tasteful interpretation of the older cultural characteristics of Polynesian islanders, whose descendants today have adopted the contemporary cultural norms of the islands. Its success is predicated on several planning and management issues: Representatives of the cultures are willing to wear the costumes and play the parts of their ancestors. The selection of archaeological period has been carefully chosen. Technicians and management staff are not visible to tourists. Finally, the design of sites and structures is carried out authentically and in a manner that skillfully exhibits and interprets characteristics of these cultures of the past (Stanton 1977).

Sustainability has been and continues to be a major policy of the U.S. National Park Service. In addition to the service's new *Guiding Principles of Sustainable Design* (1993) manual, each park is being evaluated for environmental enhancement (Strutin 1994). New recommendations have been made for New Mexico's El Malpais National Monument, San Francisco's Presidio, and Grand Canyon National Park. Proposed at the Grand Canyon is an orientation center to guide visitors into the several ways to experience the park, such as by bicycle, trail, and tour bus. Information on resources and their need for conservation is to be presented. At the Everglades National Park, the site and visitor center errors of the past will be solved by a new visitor interpre-

tation center surrounded by native sawgrass instead of a lawn and boardwalks to prevent erosion of fragile land resources. A major change is the new understanding crafted between designers and managers so that sustainability can be implemented.

CONCLUSIONS

From this review of host–guest relationships and issues of tourism development, it must be clear that tourism is multidimensional, demanding a much greater depth of understanding than is popularly held. Surely within all communities are many constituencies with different goals and agendas, but their differences do not compare to the contrasts between tourism's hosts and guests. These are two different worlds.

Because tourism is driven by two forces—demand and supply—all who are in any way involved must have thorough understanding of both. Such awareness and knowledge requires new learning in both directions. What travelers are like and what they prefer as attractions and services represents the demand side. To reach an understanding of this side, communities must view themselves as the travelers do—not an easy task. The supply side must be in response to this demand. The remainder of this book is an attempt to enlighten and give communities guidance for positive tourism development that protects resources, reaches toward community goals, and reduces negative impacts.

CHAPTER 2

Politics and Ethics

A resort operator may experience a severe drop in visitors because of a change in foreign policy. A developer may discover that his investment in coastal land is futile because a health agency has closed the beach following its contamination. Residents of a community may become incensed and begin an exodus because local policy has promoted millions of tourists who now cause congestion and social conflict. A coastal resort owner may lose millions of dollars of investments because storms and tidal waves have eroded the foundations and damaged facilities beyond repair. A government policy in a developing nation may force native peoples away from their traditional sites of livelihood into new compounds so that theirs may be converted into an industrialized and tourism society. These are policy concerns.

At the same time tourist business growth is taking place, there is also increased awareness of the role of tourism policy at all levels of government. As the negative social, environmental, and even economic impacts of tourism are being recognized to a much greater extent, new concerns over tourism policies are being expressed.

Equally significant is the recognition by observers, scholars, and activist groups that solutions to problems are not coming entirely from science or technology but also from judgments over the right-and-wrong aspects of tourism development. These are ethical issues.

In the end, it will be decisions over trade-offs among the several choices of tourism development that will set the course for the future.

POLITICS AND POLICIES

Throughout the world, all human activity, including tourism, is directed by politics. Developers of tourism must act within the canopy of existing policies. Of course, policies range widely, from highly sophisticated legal foundations to oral traditions. In democratic societies policies are the result of public input carried out through their elected leaders. In other areas, dictators, kings, tribal chiefs, and tradition direct policy. Encompassed in all are philosophies and beliefs regarding what should and should not be done and how. Policy has been a guiding force for centuries and today must be followed by investors, planners, and managers of all tourism supply development.

As observed by political scientist Richter (1989, 181), "even in developed countries tourism development has not been understood as the complex social and political phenomenon it truly is." The situation is even more complicated—and sometimes tragic—in developing nations, where economic resources and political stability are often in short supply.

Public Sector Role—National

By default, the responsibility of setting policy for tourism falls mostly on the shoulders of the public sector. The private sector is ill-equipped to deal with policy except within the narrow scope of each business. Hotel management, for example, may have established its own policies of guest service, revenue-cost ratios, target markets, and levels of maintenance. But business is, and must be, directed toward short-range profitability, especially in tourism because of the capricious nature of tourism markets. The scale of business is oriented to the site, not the destination area or region. Therefore, nations around the world have had to establish many kinds and levels of tourism policy.

Spurred by the motive of increasing economic input from foreign visitors, tourism often becomes an element of the foreign policy of nations (Edgell 1993). Although tourists are usually welcome, many regulations demand careful scrutiny of immigrants, and both must pass through customs inspection. Of concern for immigrants are the jobless, refugees from civil or international hostilities, terrorists, and those with contagious diseases. Some nations, such as the United States, have waived nonimmigrant visa requirements.

Examples of changes in national policy regarding tourism are abundant. The fall of the Berlin Wall and the Iron Curtain in Europe in 1989 opened up vast geographic areas as available tourism destinations. Cuba's policy, once denouncing tourism as a capitalistic evil, has now changed to aggressive promotion. Many developing countries in recent years have incorporated tourism stimulation and subsidies into their national policies. Alongside economic motivations have been those of enhancing political stature, fostering national identity, and cultural preservation (Richter 1989); because the resources of those nations have heretofore largely been undeveloped, they now can offer a competitive edge over older areas for the establishment of new travel business.

As Edgell (1990a, 1990b, 1993) and others have pointed out, tourism has increasingly become incorporated into recent modifications of national policy. Although tourism practices in the United States have been left mostly to the private sector, the U.S. International Travel Act of 1961, since modified, was directed toward reducing barriers, educating business on tourism matters, and promoting visitors to the country. The 1981 National Tourism Policy Act broadened policy goals, such as "to contribute to personal growth, health, education, and intercultural appreciation of the geography, history and ethnicity of the United States" and to "represent United States travel and tourism interests before international and intergovernmental meetings."

National government involvement in tourism in Scotland began with the establishment of the Scottish Tourist Board in 1969 (Scottish Tourist Board 1995a), whose purpose was to encourage development of visitor facilities and coordinate tourism interests. Its main function is promotion, but its activities in recent years have been broadened. Its Visitor Services Program includes grading and classification and a quality assurance program (Scottish Tourist Board 1995b). The Area Tourist Boards (ATB), of which there are fourteen, are given organizational and financial support. Assistance is provided in implementing the organization's Strategic Plan of 1994. Expenditures of approximately 17 million in 1994–1995 were distributed over the following divisions: marketing, convention bureaus and exhibits, visitor services, ATB grants, research, planning, finance and administration, development, and press and public relations. Clearly, the national tourism policy encompasses a variety of functions directed toward fostering tourism planning, development, and promotion.

Edgell (1990b) has cited many international organizations with strong political influence on governments regarding tourism policy, such as the Organization for Economic Cooperation and Development (OECD), the Organization of American States (OAS), the World Tourism Organization (WTO), the Committee on Capital Improvements and Invisible Transactions (CMIT), the International Civil Aviation Organization (ICAO), the

Customs Cooperation Council (CCC), and the General Agreement on Tariffs and Trade (GATT).

Nations frequently establish regulations for their own reasons that turn out to be restrictive for travel. The WTTC (1993), for example, has cited several reasons for many nations to require visas, such as preventing entry by undesirable aliens and to raise revenues. Other barriers include exit taxes and export currency control. For example, nationals leaving the Philippines pay a $7.50 (U.S.) departure tax, and travelers leaving Thailand pay $8 (U.S.). With few exceptions, Asian countries have currency limitations on those leaving their countries, especially Indonesia, Thailand, Sri Lanka, India, and Pakistan. Often, the traveler is burdened with a variety of taxes that increase travel costs appreciably, such as customs fees, agricultural inspection fees, car rental taxes, airport fees, and hotel room taxes.

Nations often look to tourism to foster their national image (Richter 1994). Pakistan's promotion focused on enhancing its international presence, creating better political directions, and counteracting bad press. The Philippines has used tourism promotion to enhance the image of martial law. Israel has promoted its image as the Holy Land to improve travel to the nation. Many nations have sought and obtained financial aid for tourism development. Tourism policy has been used to enhance impoverished areas. The Association of South-East Asian Nations (ASEAN) has been instrumental in fostering international politics of tourism, especially with Switzerland, Spain, Bermuda, and Iceland. Tourism in destination countries, however, has sometimes been negatively influenced by policies in origination countries that cite many reasons for not traveling to these areas. Other issues that affect national tourism policies are land use conflicts, inflation, urbanization, crime, and political instability.

A major policy concern of most nations is air transport because of its critical linkage between travel market origins and their destinations. Policy topics have focused primarily on safety, national security, technology, air congestion, taxes, traffic control, and computer technology (OECD 1990). In recent years, governments worldwide have been following the lead of the United States toward deregulation and privatization. In 1995 (Lipman 1995), airline administrators made many recommendations for new international air transport policies, such as better research foundations for taxation, rules of taxation equal to those for manufactured goods, greater ease of application and collection, and stronger linkage between taxation and tourist use. Many believe that existing taxes are excessive and restrictive. Because travel is an export industry, many of the present taxation policies are self-defeating.

At issue in the United States in 1995 were certain projected cuts in government spending, especially for the Department of Transportation, as part of the effort to balance the federal budget (Wright and Stoller 1995). Amtrak, the main national passenger railway, faced being eliminated entirely, and some 108 small towns serving rural areas faced losing their air service if subsidies under the department's Essential Air Service program were cut.

For many nations, policies relating to paid holidays (vacations), education, religion, and trade have indirect effect on tourism. Several Middle Eastern nations, such as Saudi Arabia and Oman, fear the potentially negative impacts of foreign visitors on their culture, especially from Western tourists. They often limit or even prohibit all entry except for diplomatic or business purposes. Although overtly the Peoples Republic of China has recently expanded its tourism interest, there remains considerable covert skepticism regarding its value. "Separate hotels, each complete with guards to make certain that Chinese and various tourist types do not mix, are a constant reminder that, whatever is happening on the economic scene, the political environment is still one of distrust and insecurity" (Richter 1989, 46).

Worldwide, there is growing need for national policies controlling the removal of ancient artifacts by tourists. The threat to the integrity of historic, prehistoric, and cultural features worsens as mass travel market demands for these features grow. When hordes of tourists spill out onto ancient castles, temples, and rare cultural sites, considerable vandalizing and pilfering occurs if there are no policies and managerial mechanisms for control. The sale of these artifacts is a huge black market economy. Especially threatened are sunken ships and their cargoes because of increased underwater technology and the popularity of scuba diving. Although some legitimate salvage operations do take place, nautical archaeologists express great concern over the loss in research findings when items are removed from their contexts and trafficked for treasure hunters. Equally vulnerable to tourist interference and destruction is underwater life.

Developers of tourism must respect conservation policies directed toward protection of endangered species of plants and animals. Beyond the emotive language of environmentalists, many nations have developed policies restricting the harvesting of certain animals, destruction of wildlife habitat, and removal of rare plants. Founded in the common weal, these policies often cause developers to discover that their chosen sites for tourism enterprises have severe restrictions or are off limits entirely. The purpose here is not to debate the issues of environmental policies but to call developer attention to the need for understanding these policies before taking action.

Although tourism is driven by market demand, some restraint of this demand may be required of developers at times. If development is allowed to proceed unchecked, there may be serious social, environmental, and economic impacts. A high demand for "native" tourism can result in degradation of the natural resource values that travelers seek. A high demand for rare animal trophies can deplete stocks of endangered species. A high demand for "travel sex" may discourage other travel segments, spread disease, and undermine traditional values locally.

One must conclude that a nation, province, or locality must consider all travel market options and make every effort to prohibit those market segments that may be detrimental. Policies are the result of social judgments and decisions, and advocacy is an essential part of politics and policies (Hall 1994). Strong policies, practices, and management controls are required for the betterment of tourism.

Dieke (1993), in his study of tourism policy for Gambia, concluded that, although it was successful on the whole, there was continuing need for better integration among several national agencies. Tourism policy there began in 1972 with the identification of potential sites and the setting of social and economic goals. Many support programs, several new agencies, and investment incentives have stimulated considerable growth (42,752 arrivals in 1979–1980; 101,419 arrivals in 1990–1991) (Dieke 1993); but some have expressed concern over the country's typical Third World pattern of external investment's control over tourism's employment and economy.

An example of new national tourism policy is the National Tourism Development Program of Lithuania (Stauskas 1995). This policy, approved by government in 1994, benefitted from the country's dramatic new political and tourist freedom following its isolation for many decades. This policy program emerged from public involvement as well as professional guidance from a university. It includes environmental as well as economic concerns. Its basic principles include:

- Nature protection has priority; tourism development is to be placed only outside protected areas.
- A qualitative approach dominates; adapt and modernize present hotels.
- Encourage hotels in towns, not in lake and seacoast areas.
- Encourage tourist flows; minimize new concentrations of development within Lithuania.
- Reorganize types of tourism to be in harmony with their settings. (Stauskas 1995, 49)

Resource analysis of the country resulted in the identification of nineteen priority areas for tourism, where natural, cultural, and historical resources are most abundant. All are accessible by car, rail, and water routes and have service communities available. To encourage quality development, the development program offers specific guidelines. Building construction is prohibited on waterfront greenbelts of 200 meters. Hotel and other building height is limited to the height of nearby hill and tree silhouettes. Indigenous plant materials are to be used for landscape planting. For tourism implementation, regulations can be enacted locally, taxes can be eased, licensing can require conditions, and the existing Law of Protected Areas must be enforced. The primary author of this program emphasizes the principle of extreme care and respect for nature in all tourism planning and development.

Certainly, both business and pleasure travel are greatly influenced by political decisions. More and more, national policies are stretching much beyond short-term economic concerns and into environmental and social controls.

Public Sector Role—Other Levels

Public sector tourism policies at the state, provincial, and local levels are becoming more important, especially in democratic societies. In European countries and other developed nations the local citizenry is gaining a stronger policy voice, both generally and specifically for tourism.

Recently, in the United States, native American tribes have been empowered to develop tourism. These "tribal nations" are now marketing their cultural assets. The Pueblos at Taos, New Mexico, the Navajos of Arizona, and the Cherokees of North Carolina have established their own policies for successful tourism development that includes scenic, historic, and ethnic attractions (Smith 1995). Residents of the Acoma site, a pueblo located about 60 miles west of the New Mexico capital of Albuquerque and only 11 miles from a major transcontinental freeway, have created their own tourism policies. The pueblo has become a major destination whose attractions include traditional ethnic festivals, skillfully designed and executed handcrafts, major restoration of homes, a museum, guide services, and a casino. "They have maximized their heritage, commodicized it without seriously altering the product" (Smith 1995, 3).

A case of inadequate community tourism policy resulted in the erosion of historic Central City, Colorado (Stokowski 1992). This small town, because of its rich mining lore and history, had been designated a National Historic Landmark. Because progress had passed it by, it still contained many of the structures and flavor of its mining heyday in the 1870s. However, the town's major opportunity was lost because of weak local policy. Recently the lure of outside casino investment overwhelmed the community's leadership with promises of historic preservation and boom economic returns. Today, little remains of the original historic mining town stores, hotels, and residences, and the economy has been captured by outsiders.

Residents report high levels of stress, traffic congestion is oppressive, open space is disappearing, historic buildings are being replaced by casinos or new housing, interiors have been gutted for gambling, and senior citizen deaths have been attributed to the loss of their homes and disruption of their lives. These negative impacts of

casinos are not necessarily always the result—only when rare resources (such as historic structures) are demolished by them. This form of intrusion can be avoided with strong tourism and environmental policies.

Taxes placed upon tourism are often used in several ways; the majority are applied to costs of promotions. Many city administrators look to tourism taxes as replacing revenues from dwindling local economic activities. Some argue that tourist taxes should be used in the general funds of cities because tourism increases demand for public services such as water supply, sewage and waste disposal, and fire and police protection.

Often tourism divides a community between those supportive and those critical. A case in point is the small town of Niagara-on-the-Lake, Ontario, a village that played an important role in the American War of 1812 (Rumble 1994; Stokes 1994; Howe 1994). Its main street, with its restored historic buildings with modern shops, and the nearby Shaw Festival Theater attract approximately 3.5 million visitors annually, mostly in two summer months. A political issue today is whether further growth should be limited. Even though tourism provides an economy that supports attractive and well-maintained civic amenities, many local citizens are disgruntled. They now must shop elsewhere for basics because most businesses have been converted to tourist trade. "Trippers" (short-term visitors) clog the streets, buses pollute the air and quiet, and older residents nearby are selling their homes because of the erosion of their quality of life. Town council members and provincial leaders now face serious political decisions regarding the future.

In Third World countries, *enclave tourism* has been a dominant pattern, a clustering virtually isolated from the local people. This pattern is usually created and managed by outside investors. Local policies that encourage such development generally have produced poor local results because the power and economic advantages occur elsewhere (Lea 1988). Gradually, however, with new understanding of the consequences, mega-resort development in these countries is being supplemented by small-scale, adaptive, and localized tourism.

Tourism development policy at the local level has varied from nonexistent to sporadic and variable. Although it is difficult to speculate on the reasons, general observation suggests this poor record is because of a poor understanding of the complexities of tourism, a lack of local leadership, and a lack of enabling legislation. Without any guidance by the local people, outside investors frequently take advantage of their gullibility and create conflicts with local lifestyles and land uses, especially resource protection.

An example in which a state government has taken the initiative to solve this issue is Arizona (Andereck 1995). Following a governor's conference on rural development, a special program was recommended and is now identified as the Arizona Council for Enhancing Recreation and Tourism. For the first time, many public and private entities are being coordinated toward the goal of planned tourism development. The program is demonstrating, in its first application in the community of Globe-Miami, that public policy at the state level can be an effective mechanism for improved tourism development. Other pilot communities include Hualapai Indian Reservation, Douglas, Parker, and a Hopi reservation called First Mesa Village (Leyva, 1996).

Public–Private Policy Roles

Generally, urban planning agencies and professionals have not viewed tourism as within their scope of interest and responsibility. Virtually all urban planning theories and practice have been directed toward residents. Today, with burgeoning mass tourism being focused heavily on cities, the narrow resident approach is insufficient. Badly needed are urban planning policies adapted to both visitors and citizens. As Ashworth and Tunbridge (1990) have pointed out, there is both overlap and conflict between tourist and resident in cities. The resident function is repetitive and familiar; that of the tourist is transitory and new. The resident has a more comprehensive view of the city and all its amenities; the tourist sees only those attractions programmed into his itinerary. Even the independent traveler generally visits only the "touristy" features of a city. Planning policies are needed that recognize the spatial and management needs of visitors as well as citizens, especially regarding transportation, shopping areas, historic areas, tourist services, and the many interrelationships among them.

Increasingly, the polarized standoff and antagonism between the public and private sectors of tourism is being seen as inhibiting progress. An example of one solution is a new nonprofit organization in Ontario, Canada (Hawkins 1994). Based on a government initiative, regular meetings between the provincial agency and tourism representatives resulted in *Ontario's Tourism Industry: Opportunity, Progress, Innovation* (Canada 1994). A major conclusion of this report was that there was a lack of any formal organization to follow through on the recommendations. Because of this conclusion, the Ontario Tourism Council, made up of tourism industry representatives, was formed August 23, 1994. Its core responsibilities include:

- providing an effective mechanism for increased communication and cooperation within the industry so that partnerships are encouraged and duplication of effort is diminished;
- overseeing the effective implementation of the Tourism Strategy and working to resolve issues or barriers that may arise;

- assisting in the formation of the marketing and advocacy organizations and the strengthening of the Ontario Tourism Education Council (OTEC);
- ensuring that sound business principles are used based on best practices; and
- influencing all levels of government on tourism-related major capital developments and infrastructure in Ontario. (Hawkins 1994, 119)

The province of Alberta has likewise established a new partnership between the public and private sectors of tourism (Alberta Tourism Partnership 1996). On February 25, 1996, a contract was signed between the newly formed Alberta Tourism Partnership Corporation and the government of Alberta. This agreement transferred the responsibility of marketing and coordination of tourism services from the government to the business sector. This action followed many others, such as the establishment of the Tourism Alliance for Western and Northern Canada, with offices in Saskatoon, Saskatchewan, and Vancouver, British Columbia. Strong support came from the Tourism Industry Association of Alberta (TIAALTA).

Canada has a long history of stimulating provincial and local policies on tourism development. For example, the province of Alberta's *Tourism 2000: A Vision for the Future* (Tourism Alberta 1991) cites the success of a partnership program between the provincial agency and the private-sector TIAALTA. Funding assistance is provided to communities to develop community tourism action plans. Such plans set goals, identify local projects, and establish programs as well as stimulate interest and enthusiasm for new tourism opportunities.

Theoretical foundations for greater cooperation and even collaboration for tourism development at the community level have been identified. Put forth by Jamal and Getz (1995) are six propositions directed toward manager, planners, and researchers: (1) a recognition of the high degree of interdependence of all elements of tourism; (2) a recognition of mutual benefits possible from cooperation; (3) an anticipation that results will be carried out; (4) an understanding that key stakeholder groups are involved in the process; (5) an understanding that a respected mediator will facilitate cooperation; and (6) a clear identification of vision, goals, objectives will be made. The projected outcome of these propositions is environmentally sound and socially adaptable tourism development. The purpose is to maximize business success and avoid common pitfalls of joint ventures and less suitable approaches.

Certainly, no designer, planner, developer, nor tourism advocate can escape the influence of public policy. Policies can inhibit, foster, or control tourism. An awareness of existing policies and how they can be modified for best adaptation of tourism to localities and nations is essential.

ETHICS

The driving force behind all business, tourism included, is profit making. This objective must be understood as encompassing not only returns to the investor but also the many costs of doing business. However, today the profit-making drive is being modified by ethics. This change may even be underscored by law, but it derives mainly from beliefs that govern behavior, especially on a broad societal level. Petulla (1980, 209) has stated:

> Gradually, then, ethical norms become more general-
> ized, more socialized, so that they rest on a common
> perception of the truth accepted by a large body of
> people who share a common conception of right and
> wrong.

Most of tourism's great growth can be laid at the door of new technology—spectacular changes in air transport, major upgrading of automobiles and highways, computer technology, air conditioning, improved building construction techniques, and creation of new fabrics and clothing materials. Viewed by both undeveloped and developed nations as progress, these changes have been adopted as rapidly as possible. As a result, mass tourism has exploded in volume throughout the world as a desired good without anyone asking the question "Should it be done?" This is neither a scientific nor a technical question; it is a moral and ethical one. Observed Shepard (1967, xviii), "The headlong race into a mechanical cornucopia conceived on the pushing and hauling of the physical foundations of our environment is worse than military brinksmanship because we have fewer moral and ethical restraints on our disposition of nature than of each other."

Today, local people and even governments are beginning to recognize that there must be opportunities for examining tourism's potential more carefully and, in some cases, denying its development, solely on the basis of judgment. More and more, it is being demonstrated that there are right and wrong ways of developing tourism and that those people who would be affected have a right to make these moral decisions. Today's moral questions are based primarily on new revelations of potentially negative social, environmental, and economic impacts. Heretofore, tourism was considered a free good with nothing but positive economic results. In recent times, for example, Giltmier (1991, 43) stated, based on Aldo Leopold's (1949) philosophy:

> Meanwhile, we should quit thinking about land use solely
> as an economic problem. We must examine each ques-
> tion in terms of what is ethically and aesthetically right,
> as well as what is economically expedient. A thing is
> right when it tends to preserve the integrity, stability, and
> beauty of a biotic community.

Lea (1993) has summarized the existing literature on tourism ethics to include three general categories: development in the Third World, social and physical impacts, and traveler behavior. His study found that calls for changing tourism development policies are being heard from many sources based on ethical grounds. Some are calling for rigid control and others for complete prohibition of visitors, even on a global scale.

Tourism ethics is now seen as a topic of research and implementation (see Appendixes). The following brief discussion introduces communities, planners, and developers to the growing need for concern over environmental, social, and economic ethics.

Environmental Ethics

The term *environmental ethics* has its roots in Aldo Leopold's (1949) land ethic, which elevated the rights of plants, animals, and all nature to the same level as those of humans. Although arguments over environmental ethical issues are sometimes supported by research facts, the concept is primarily one of belief, even approaching religion (Petulla 1980). Although often attacked by economic realists—such as those who would harvest timber in rain forests—decisions are usually on the basis of belief in what is right and wrong for the majority of society and for the long range.

More than a century ago, in 1872, the decision to dedicate the land of Yellowstone National Park to the public rather than the private domain was a moral decision. It was decided that, because of the very special land attributes of the area, it would be wrong to allow individual property rights to control it—a larger social ethic prevailed. For the good of society, a dual mandate was established: resource protection and public use by visitors. In no way did this decision erode the basic U.S. policy of supporting private rights of land use generally. But it did illustrate an application of ethics for the development of tourism.

From the environmental perspective, among the studies and principles being put forth, perhaps the *Eco Ethics of Tourism Development*, published by The Western Australia Tourism Commission (1989), may be said to offer the best guidance. This state tourism agency, in collaboration with the Environmental Protection Authority and the consulting firm of Brian J. O'Brien and Associates, has prepared succinct statements on the right and wrong ways of developing tourism. Its definition of ecoethics follows:

> Moral principles or values dealing with behavior that has,
> or could have, environmental impact, implications or sig-
> nificance, the principles of environmental conduct gov-
> erning an individual or a group in any interaction with
> the biosphere. (Western Australia Tourism Commission
> 1989, 30)

Put forth in this document are principles for tourism development that recognize the great value of the environment to tourism and therefore the need for interaction in a positive rather than a destructive manner. These ethical principles are grouped into eight headings: all environments, beaches and ocean frontage, remote environments, forests and national parks, pastoral leases, waterways and wetlands, heritage (the built environment), and aboriginal culture. Because the report's authors knew that implementation of a code of ethics might be difficult, specific strategies are recommended. Implied is a strong role by government, especially through the Australian State Conservation Strategy of 1987.

The issue of environmental ethics for tourism has also been recognized by the private sector. For example, the WTTC in 1991 established its research center, WTTERC. In 1992 it published its first annual review, and in 1993 the second annual review (WTTERC 1993, 6) addressed practical aspects of achieving environmental ethics in tourism development. Stressing the responsibility of business to take leadership, the report made the following statements:

- Travel and tourism are integral aspects of modern societies.
- Global awareness of environmental damage is developing rapidly.
- The resources of the world's largest industry can and must be harnessed to achieve environmental goals.
- The industry has the potential to influence billions of customers per year and to use its leverage to achieve beneficial environmental effects.
- The customer challenge will exert a growing pressure to achieve environmental improvements.
- Self-regulation must be developed rapidly and effectively and used to influence the development of appropriate and workable regulation.
- Corporate environmental mission statements are a vital first step toward self-regulation.
- Environmental leadership must come from the major international companies.

These ethical challenges certainly show new awareness and a call for action by the private sector of tourism. As of 1993, WTTERC held files on 65 nonregulatory (nongovernmental) agreements that affected the environmental practices of companies and the behavior of visitors.

Increased sensitivity to environmental issues is also showing up within local citizenry. For example, originating in Germany and now active throughout Europe, "eco-counselors," mostly women, counsel individuals, businesses, and communities in sustainable development principles and actions (Mudrick 1991). Usually college graduates, in France and Belgium they receive an extra one-year training program at the Eco-Conseil Institutes of Strasbourg and Namur. This program provides them with extra courses in communication, negotiation, and mediation; legal, scientific, and technical aspects of environmental matters; and information on municipal and regional administration or business management.

Another sign of the increasing concern over environmental ethics is the work of the World Bank. In 1993, the bank's environment department updated its *Environmental Assessment Sourcebook* (World Bank 1993). All tourism development projects submitted to the World Bank for financial aid are now subject to the department's regulations and judgment regarding environmental impacts. Emphasis is placed on concerns such as disturbance of tropical forests, conversion of wetlands, potential adverse effects on protected areas or sites, encroachment on lands or rights of indigenous peoples or other vulnerable minorities, involuntary issues, and toxic waste disposal. Many of these issues relate to tourism. An example is the World Bank's loan of $130 million (U.S.) to the Tourism Development Authority of Egypt to improve private-sector tourism policy, support infrastructure for two model integrated master plan developments in the Hurgada-Safaga area, and upgrade water supply, sewerage and solid waste for the existing resorts and stimulate cooperation to protect the coastal resources.

A study of local response to tourism development in the Columbia River Gorge region of Oregon revealed differences of attitudes among segments of the populations (Lankford, 1994). Residents' attitudes differed significantly from business owners, government employees, and civic leaders. Generally, residents were skeptical. They doubted the ability of government to control growth and felt that environmental and social impacts are serious concerns. Businesses supported increased development and promotion of tourism, more than any other group. Government employees and civic leaders gave mild support to tourism, believing in increased economic potential. These differences of perceptions suggest that a major moral issue must be addressed by the entire community: Should tourism be developed? All local constituencies anticipating the development of tourism must obtain a clear understanding of the environmental ethics involved.

Social Ethics

Equally significant is concern over social ethics of tourism development. It is well documented today that cultural contact creates negative effects as well as positive impacts such as cross-cultural understanding and world peace. It is essential for host areas to understand that the traveler is a stranger who often has an entirely different cultural background. Equally significant is the reverse, that the tourist understand that his objectives are far different from those of his host.

Examples of unethical behavior of tourists as they travel are abundant. Many were identified by Richter (1989) in her study of tourism policies in Asia. For example, Asian countries are increasingly disturbed by the presumption among tourists that their region is one big brothel. Elsewhere, tourists often disrespect property by invading the privacy of homes and littering their yards. Different styles of dress (or undress), modes of conversation, and levels of affluence often cause considerable irritation and conflict between tourists and hosts. The camera- and camcorder-toting tourist often uses no restraint in interrupting religious ceremonies, ethnic festivals, and weddings. When visiting foreign restaurants, American tourists often complain about indigenous foods and service and ask for directions to the nearest McDonalds or Burger King. Long ago, writer Pritchett (1964, p. 1) titled a book *The Offensive Traveler,* and stated, "By being offensive I mean that I travel, therefore I offend. I represent the ancient enemy of all communities: the stranger. . . . We are looking in on the private life of another people, a life which is entirely their business, with an eye that, however friendly it may be, is alien. We are seeing people as they do not see themselves."

But not all local responses to tourism show negative influence. For example, a study in Fiji (King, Pizam, and Milman 1993) showed overwhelming positive social adaptation to mass tourism. A cross-section of the population of Nadi was surveyed and showed a positive attitude toward tourism based on examination of the variables of impacts. They identified many improvements that tourism had brought them, including a better standard of living, work attitudes, and quality of life. Many local hosts have established continuing correspondence with tourists. Although some negative social impacts were recognized (sex, crime, drugs), in many cases the positive values far outweighed them.

Edgell (1990b) has cited a code of ethics for tourists as prepared by Ron O'Grady (1980) for the Christian Conference of Asia (see Appendix A).

Investors, planners, and communities must be aware of the potential social conflicts between visitors and local populations so that they may strive to avoid or correct them.

Business and Economic Ethics

Although the great majority of tourist businesses operate ethically, the popular travel literature is filled with cautions against unethical practices. It seems that the great asset of free enterprise in providing the needed travel services, products, and facilities is often tainted by unscrupulous operations. The ombudsman feature of *Conde Nast Traveler,* for example, regularly cites cases of unethical practices and intervenes on behalf of the traveler. Tourists to seasonal resort areas often experi-

ence the unethical business practice of jacking up rates of food, lodging, and products. Alert travelers soon get wise to this practice and take their travels elsewhere. To guard against this practice, the European Union Package Travel Directive became law in the United Kingdom in 1992 (Scottish Tourist Board n.d.). This law requires all tourist package operators to provide full and accurate descriptions of packages (combined transport, accommodation, related services) and guarantees of advance booking. It describes the ethics of brochures, contracts, and financial security such as bonding, insurance, and trust accounts.

Perhaps the greatest unethical practice within tourism is the frequent lack of congruence between promotion and product reality. The reproduction of images and the elaborate descriptions of attractions are so facile today that advertising images soar way beyond the actualities of place. For example, suppose an advertising firm given the charge of creating an ad sends a photographer to a European castle. The wait for a clear day may take weeks, and the most dramatic view requires permission from the landowner to climb to the best vantage point. The resulting spectacular photograph in the ad is in considerable contrast to the typical ground level tourism view, which is most frequently on a rainy day. Too often anticipations far exceed the reality of the traveler experience, resulting in disappointment. Instead, businesses should always strive for a visitor experience that exceeds all preconceived images. Perhaps the ultimate in honest promotion is the following excerpt from a travel guide to Pattaya, Thailand: "Pattaya's bars are what most people come for rather than its beach, which is brown and grubby" (*Thailand* 1993).

A similar concern arises concerning zealous developers of tourism. In the past there has been a tendency for promoters of development projects to omit potential economic costs in their arguments for tourism. This is unfair. Many countries, provinces, and communities are now seeking tourism for the first time because other economic foundations have weakened or disappeared. They can easily be enticed to invest in resorts, attractions, or other developments when the developers neglect to raise important environmental, social, and economic issues. These new investments may overtax the local water supply, transportation system, and police. Because outside investment often brings its own staff, little economic impact and employment accrue locally. Because it wasn't planned properly and its impacts were not considered, such tourism becomes a detriment rather than an asset. This is an unethical practice on the part of the developers.

Nelson (1991) has made a plea for a stronger cooperative ethic for tourism development, citing the English national park system, in which large areas are privately owned, and which requires local governments, tourism developers, residents, park managers, and even tourists to work together. In many instances in Canada and the

United States, tourism sectors and park agencies have collaborated on projects. Because tourism varies so greatly in scale and adaptability to local conditions, the best cooperative private–public sector development ethic is at the local level. Only by means of information, communication, monitoring, and adapting can the best solutions be derived. As a stimulant for such action in Canada, the Tourism Industry Association of Canada in 1989 challenged its Environmental Committee to develop an Environmental Code for Tourism (Lawson 1991). The objective of the project was to develop a code of ethics for sustainable tourism (statement of principles) and codes of practice for stakeholders such as tourists, governments, associations, and, most importantly, the key industry sectors: accommodation, food and beverage, and tour operators (see Appendix B).

Also in Canada, British Columbia has documented the need for codes of ethics and made many constructive recommendations in its *Developing a Code of Ethics: British Columbia's Tourism Industry* (ARA Consulting Group 1991). This report recommends that a code should address topics such as: product quality, safety, marketing, adherence to laws and regulations, organization and individual responsibility, and penalties for nonadherence. It suggests that the concern over tourism ethics has arisen from a more demanding public, pressure on wild land for ecotourism, and greater communication and debate among resource management agencies.

Throughout the world, tourism businesses are practicing positive ethical standards within their operations. Although these ethics are derived from a multitude of sources influencing employees and management, they are all directed toward the right ways of providing service to travelers. Addressing theories of ethics, Upchurch and Rohland (1995) have applied them to lodging businesses, using a classification of three: principle, benevolence, and egoism. *Principle* means that actions are influenced by ethical rules, established by either management or personality of employees. *Benevolence* refers to the overall ethical climate and that ethics guide toward the best interest of an organization. *Egoism* refers to ethics directed toward self-interest, whether for the individual or the firm. The desirable management pattern balances all three by means of organization, training, and performance control. From their study in Missouri, the authors recommended that for better understanding of ethics in the tourist business workplace, managers need to clearly comprehend the field of ethics and how application can contribute to their success. Others, such

as Przeclawski (1995) have expanded ethics for tourism to include tourists, the residents of destination areas, and all those who provide service to visitors; all influence negative or positive behavior.

CONCLUSIONS

Rather than all bad or all good, it is clear that tourism brings with it both kinds of consequences. The best development of tourism is in the balance between seeking its positive effects and avoiding its negative effects, attained through use of good planning. Decisions involve trade-offs between the potential benefits to the locality and nation and the costs (environmental, social, economic) that tourism expansion may demand. Unfortunately, the field of tourism and the understanding of its intricacies are so new that the foundation for making choices is not often clear. As Richter (1989) has pointed out, political debate over tourism often comes too late in the implementation process and only when major issues have become apparent. It is therefore critical for local areas to do everything they can to understand potential negative and positive impacts very early in any tourism development process. Only in this way can they plan, create policies, and make trade-off decisions. In the planning stage, not after construction begins, decisions can be made on predictable consequences that may damage the environment, disturb society, or increase costs beyond feasibility.

No investor, planner, developer, or community can escape the many political and ethical issues inherent in the present escalating growth of tourism. Ethical decisions have been a part of all societies for centuries and predictably will continue to guide the future of tourism everywhere.

The conservation ethic now penetrating every corner of the world was given its greatest impetus in tourism planning in 1872 with the moral and ethical decision to create a publicly owned and managed area at Yellowstone Park. This judgment has provided a precedent for creating a balance between public and private use of resources.

Observation and study of tourism trends have convinced the author that the three sectors of tourism development (governments, nonprofit organizations, private enterprise), at all levels from national to site, will benefit not only themselves but also the entire world of tourism by implementing a set of development policies and ethics.

Tourism Function: Demand

A first step in development for a community, destination, or region is to understand that tourism is driven by two forces—travel *demand* (by those interested in and able to travel), called the travel market, and developed *supply* (all the physical development and programs for tourists). For tourism to function properly, there must be a reasonable balance between demand and supply. However, it cannot be inferred that demand and supply are static and easily identified. Instead, they are extremely dynamic and complicated, requiring special study and understanding of current conditions and trends by all developers. The intimate relationship between demand and supply can be modeled as illustrated in Figure 3-1. The arrows dramatize the relationship; demand influences supply development, and supply influences markets. No matter the scale of development—from the site, community, or destination to the region or nation—this fundamental relationship is the foundation for all tourism.

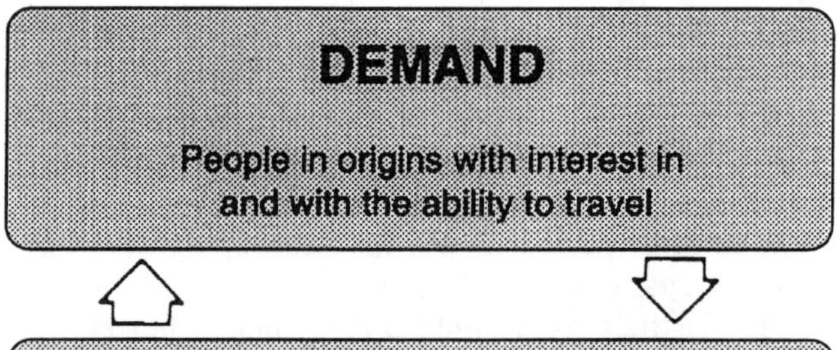

Figure 3-1. *Travel demand related to supply. Tourism development is driven by both demand and supply. Each one influences the other. Proper development strives for a balance between the two.*

TRAVEL DEMAND CHARACTERISTICS

Communities and destinations contemplating new tourism development must have an understanding of the present and potential visitor characteristics if they are to avoid the pitfalls described in Chapter 1. Although the need for this understanding is obvious, the task of fulfilling it is not simple (Pearce 1989). In spite of a plethora of travel market research in the last decade, the resulting information generally has not been structured toward guidelines easily applicable locally. The process of creating guidelines for tourism is much more complicated than for a local shop, for example. A local retail market area is easily circumscribed around a community, whereas travel markets are scattered all over the globe. Following is a brief summary of market considerations of greatest importance to local developers of tourism.

Definition

In spite of market research findings, a clear-cut definition of travelers has not yet been found. For community development purposes, a precise definition may not be necessary. However, it is difficult to know what should be developed without some understanding of travel markets.

Recently the WTO has created new definitions (WTO 1992). The term *visitor* includes all kinds of travelers, such as those visiting friends or relatives, traveling for business, making pilgrimages, and pursuing other leisure activities. Earlier definitions by statisticians usually placed a time or distance limit; the new definition includes day trips as well as longer stays. The only exclusions are commuters, soldiers, diplomats, refugees, and those who cross an international border and are paid for work within the country of the visit. The main conclusion about today's definition is that the term tourist includes nearly all kinds of travelers, for both business and pleasure.

Defining tourists in this way is a dramatic deviation from the popular image of the past—the vacation (holiday) traveler with a loud shirt (and mouth), visor cap, and camera strapped to his neck. The pleasure tourist on holiday remains a large segment of the market, but, from the standpoint of development, many other segments fill the ranks of travelers. The new definition greatly broadens its scope.

Recent technology has improved the ability of nations to identify demand and its characteristics. The "Satellite Account" used by Canada (Meis and Lapierre 1995) is a comprehensive multilayered information system that collects, orders, and interrelates statistics describing all aspects of tourism. It identifies origins, service purchases, and other traveler transactions. For modern community development of tourism, a tourist is defined as a traveler, and tourism encompasses day travel as well as longer trips, for virtually all motives and purposes.

Origin

Most travelers come from cities rather than rural areas. The geographic distribution of cities provides a clue to present and potential origins of travelers. The main relationship between destinations and origins is that of time and distance—a function of transportation. Within the United States, for example, because of its large domestic population, good roads, and high automobile ownership rate, the greatest volume of visitors travel within a radius of 200 miles from home.

However, because of competitive air fares, long distance travel to major attractions such as Walt Disney World is significant. Hawaii's tourism boom began at the start of the jet age. For other nations, international travel is very important. The points of origin of the majority of travelers to Taiwan are Asia, North and South America, and Europe (Taiwan Tourism Bureau 1995). The origins of travelers to Austria are primarily the Netherlands, Germany, Italy, the United Kingdom, the Scandinavian countries, and others nearby (Ritchie and Hawkins 1993). To Australia, most travelers come from Japan and other Asian countries with expanding economies, such as Taiwan, Korea, Thailand, and Malaysia. The significance of origin lies not only in the time and distance factors but also in the characteristics (interests, habits) of the population within that origin, especially for planning of development at a destination. Today, many tourism agencies and organizations are maintaining statistics on traveler origins. Certainly all past studies of the geographic distribution of traveler origins should be studied for clues to the characteristics of potential travelers.

Socioeconomics

Even if large populations live within a reasonable distance of a destination, they may not have the financial ability to travel or the tradition of travel. Business and industry must be profitable enough to support travel. Individuals must have a threshold of income sufficient to support travel. This is not to say, however, that travel is restricted to the affluent; travel motivation has become so strong that it is often given a higher expenditure priority than the purchases of hard goods, even for those with modest incomes.

Other important factors are the age of potential travelers and their time available for travel. The recent major increase in life span has provided a great mass of older but physically and mentally active travelers from many origins. Today, people of all ages seem to have a desire for the rewards of travel. Teenagers, as they try to break from parental control, generally do not like to travel with parents; but school and college students often participate in travel on holidays. And, as the WTO (1983) has pointed out, the time factor is important among populations at origins—biological time, work time, obligated time, and free time. All these factors vary considerably within populations. Trends suggest that in the future many more people will be traveling because of improved socioeconomic conditions within many nations.

Policy Constraints

As has been pointed out in the chapter on politics, nations vary in their policies regarding their citizens' travel. For example, Edgell (1990) has pointed out that taxes are sometimes levied to discourage outbound travel in favor of domestic travel. Indonesia in 1982 increased its exit tax sixfold for every citizen leaving the country. Some countries have made the process of obtaining travel visas so difficult and tied up in red tape that it is almost impossible to leave the country. Some national policies restrict their citizens' travel to only those countries whose culture, religion, and political systems are similar to those at home. In several nations, only business travelers are admitted for fear that pleasure travelers will corrupt the mores of their citizens. Even though many constraints on travel at points of origin have been reduced in recent years, this is a factor to be considered in evaluating tourist potential.

A community that is planning to expand its tourism should have some understanding of potential visitors, especially their geographic origin, their socioeconomic status, and policy constraints on them.

CLASSIFICATIONS

Interests and Activities

Perhaps the most crucial characteristics of travelers are their preferred interests and activities. Market research has revealed how important travel interests are in selecting destinations. A proponent of market–plant (supply) match in Canada, Taylor (1980) performed market research in several countries. This research revealed that several of the destination preferences of people in potential countries of origin, such as warm sunny beaches, could not be offered easily in Canada. He found that, of the six segments of Swedish travelers, the Canadian plant could satisfy only one. German markets were matched by only two. All American markets, however, matched the available Canadian plants.

Many attempts have been made to segment travelers by their activity interests. Crompton (1979) grouped pleasure vacation travelers by their motivation: escape from a perceived mundane environment, exploration and evaluation of self, relaxation, prestige, regression (less constrained behavior), enhancement of kinship relationships, and facilitation of social interaction. Shoemaker (1994, 9) segmented travelers based on Lundberg's (1971) grouping by motivation:

Educational and cultural motives: to see how people in other cultures live, work, and play; to see particular sights; to gain a better understanding of what goes on in the news; to attend special events; to participate in history (visit temples or ruins) or in current events.

Relaxation and pleasure: to get away from the everyday routine, to have a good time, to achieve some sort of sexual or romantic experience.

Ethnic: to visit places one's family came from, to visit places one's family or friends have gone.

Other: weather (for instance, to avoid winter), health (sun, dry climate), sports (e.g., swim, ski, fish, sail), economy (inexpensive living), adventure (new areas, people, experiences), one-upmanship (keeping up with the Joneses), conformity, sociological motives.

Travel market segmentation must be considered in all tourism planning. Some prefer active and physically challenging activities at destinations. (Photo courtesy Carson Watt)

Other travel segments choose destinations where they can be enriched by cultural assets, such as the great musical heritage of Austria and Vienna and a visit to the Johann Strauss monument.

His research generalized vacation travelers into three categories: get away/family travelers, adventurous/educational travelers, and gamblers/fun travelers.

However, care must be exercised in assuming that segments are homogeneous in characteristics. For example, Graef (1977) found that tourists taking trips down the Rio Grande through Big Bend National Park did not report a single benefit from the experience. Following is the variability found within this segment: learning about nature; enjoyment of solitude and freedom from stress; finding a challenge or adventure; increased self-awareness, learning more about oneself; improvement of status in the eyes of others; sociability or companionship with other people; enjoyment, having a good time; autonomy, feeling independent and doing things on one's own.

A very general way to classify travel markets, especially useful for tourism development, is to group them into two categories based on activities: those associated with touring circuits and those linked to longer-stay destinations. These are listed in Table 3-1 (Gunn 1988, 92).

A study in Australia divided Australian pleasure travelers into six segments (adapted from Hollinshead 1993, 641):

1. *New enthusiasts* (16%): interested in experimenting, fantasizing, and visiting different cultures.
2. *Big spenders* (18%): passive, seeking luxury and group travel: conformists.
3. *Anti-tourists* (18%): nonconformists, seeking individual experiences of quality and authenticity.
4. *Stay-at-homes* (11%): prefer planned trips, motor-coach tours, no cultural contrast or new adventure.
5. *New indulgers* (14%): similar to big spenders, but with emphasis on rich and colorful experiences, pacesetting, trendsetting.

TABLE 3-1. Activities Classified by Length of Stay

Touring circuit activities categories	Longer-stay activities categories
Driving for pleasure, sightseeing	Vacationing at resorts (food, lodging, fitness, recreation)
Visiting outstanding natural areas: parks, forests, scenery	Vacationing at camp sites: parks, forest areas
Travel camping: tent, trailer, recreational vehicle	Vacationing at hunting, fishing, other sports destinations
Water touring: boating, cruising, rafting	Participating in programs at organization camps
Visiting friends or relatives, including duty travel	Visiting personal vacation homes
Visiting universities, factories, processing plants, science facilities	Participating in festivals, events, pilgrimages
Visiting national and state shrines, pilgrimages, gardens	Participating in conferences, conventions: professional, business
Visiting places noted for food, entertainment	Vacationing at gaming centers: gambling, racing, entertainment
Visiting historic or archaeological sites, buildings, museums	Visiting major sports arenas: domes, coliseums
Visiting places important for ethnic foods, costumes, arts, drama	Visiting major trade centers: professional, business
Visiting shopping areas	Visiting science-technology centers: professional, business
Visiting art, craft, gift, or legendary places	Vacationing at theme parks

Source: Gunn 1988, 42.

6. *Dedicated Aussies* (22%): domestic interest, nationalistic, congregate in their own groups, dislike experimentation.

This sampling of travel market characteristics demonstrates the variables that need to be considered when communities and destinations begin to plan for tourism development.

Ecotourists

In recent years, the topic of ecotourism has emerged. Generally, ecotourism has come to identify the markets and support systems related to conservation and natural resources. Many scholars have attempted to define ecotourists and ecotourism. Wight (1996, 1) has suggested that ecotourist markets involve:

- experiencing, respecting, and understanding resources;
- undertaking the experience in an environmentally aware manner;

- using support services or facilities that are environmentally sensitive; and
- contributing directly to local economies.

Such a comprehensive definition bridges both the supply and market sides. The traveler's interests in ecotourism activities will vary depending upon the characteristics of the destination. But, on the basis of her research, Wight (1995, 12) has listed the following reasons (in rank order) why an ecotourist trip has appeal:

- enjoy scenery/nature,
- new experiences/places,
- land activities,
- wildlife viewing,
- see mountains,
- wilderness experience,
- not touristy/crowded,
- water activities,
- cultural attractions,
- study/learn nature/cultures,
- rest/relax/get away,
- been there, go again.

Research into ecotourism markets (Wight 1995, Pearce and Wilson 1995) has revealed that interests are quite varied from individual to individual, especially among international travelers who prefer multistop opportunities. Ecotourism markets for Canadian destinations have been ranked as follows: visiting a wilderness setting, wildlife viewing, hiking/trekking, visiting national parks/protected areas, rafting/canoeing/kayaking, casual walking, learning about cultures, and participating in physically challenging programs.

Ecotourism, once thought to be a fleeting fashion, is proving to be a firm and predictably growing travel segment. WTO has predicted that by the year 2000 most of the 86% increase in tourism receipts will have come from active, nature- and culture-based travel (Reingold 1993). But ecotourism, more than most other forms of tourism, places natural resources in jeopardy. One example is the controversy over the ten-day multi-outdoor sport Eco-Challenge, which traverses 370 miles of Utah's wild lands (Glick 1995). The Southern Utah Wilderness Alliance and the Wilderness Society argue that this event exacts serious destruction of fragile desert flora. Much of the controversy stems from a lack of environmental study and planning before the initiation of the event.

Often, it is tourists themselves who express concern over management practices that allow destruction of resources. A study of tourism in the New Zealand sub-Antarctic islands (Cessford and Dingwall 1994) showed that tourists readily understood and accepted the regulations required to protect the environment and control visitor use.

A task force study in Texas (Task Force on Texas Nature Tourism 1995, 2) identified "nature tourism" as "discretionary travel to natural areas that conserve the environmental, social and cultural values while generating an economic benefit to the local community." It showed hunting to be worth $1.07 billion, but nonconsumptive outdoor recreation such as bird watching, nature study and photography, backpacking, hiking, boating, camping, rafting, biking, and climbing to be of even greater value. Wildlife appreciation participants in the state increased 61 percent between 1980 and 1995. The American Birding Association was quoted as listing Texas as the most popular destination in the United States for birding tours over the previous five years. It was predicted that by the year 2000, 18 million Texans would participate in nature tourism.

Cultural Tourists

The cultural aspect of ecotourism is often given special emphasis. Learning about another culture is often cited as a strong travel motivation (HLA Consultants and ARA Consulting Group 1994). The (U.S.) President's Committee on the Arts and the Humanities has recommended the "Protection, preservation, enhancement and development of local and regional cultural tourism assets. These assets include historic sites, buildings and heritage areas; natural areas, scenic routes and vistas; facilities and geographic areas that provide the setting for cultural expression and actual art forms and traditions" (Moskin and Guettler 1995, 21).

The cultural tourist represents a comparatively recent market segment in the United States, whereas tourists

National parks and preserves provide for traveler activities associated with natural resource settings, such as wildlife, forests, mountains, and lakes. Planning and management must ensure sustainability. (Photo of Yellowstone National Park by William S. Keller, courtesy National Park Service)

The travel segment interested in cultural and historic resources, such as Asia's rich past, continues to grow throughout the world. Visitors now can learn about the lifestyles and philosophies of Thailand at the Phra Nakorn Khiri Historic Area Park.

Tourist interest in ancient civilizations, such as the Maya civilization that dominated portions of Central America from about 200 to 800 A.D., continues to grow. Tikal National Park, Guatemala, illustrates the challenge of archaeological restoration and controlled visitor use.

have been visiting Europe for its cultural attractions for generations. Until recently, critics did not believe the nation was old enough to value its past. Within the nation, the drive toward business, industry, and progress was so strong that many citizens saw no reason to protect older buildings and artifacts. Today, the climate for archeological and historical preservation has reversed dramatically. Many public agencies and nonprofit organizations have begun to cater to the cultural tourist segment of ecotourism.

Historian Tighe (1990) has described the cultural tourist as one who experiences historic sites, monuments, and buildings; visits museums and galleries; attends concerts and the performing arts; and is interested in experiencing the culture of the destination. Table 3-2, included in Tighe's presentation, illustrates foreign visitor participation in cultural tourism in the United States. Domestic interest in culture-related activities is equally strong, as shown by a survey of the U.S. population in 1983 (Survey Research Center 1983): 22% had visited an art museum, 17% had attended a musical, 13% had attended a classical event, and 12% had attended a play.

An important travel segment involves visiting friends and relatives while traveling. This is perhaps the most significant way to experience the culture of the destination that surrounds the friends or relatives; as the traveler makes the visit, he is often given insight into the local milieu, occasions to meet local people, and the opportunity to penetrate the local culture in much greater depth than usually offered by tours. The local host thus plays a significant tour guide role, much superior to that of guides in most organized tourism.

Among the many activities sought by travelers, the search for and the purchase of souvenirs have formed a significant segment of the tourism market. Littrell et al. (1994) have documented many studies revealing the shopping interests and habits relating to souvenirs as symbolic evidence of the travel experience. Their survey of travelers to Iowa, Minnesota, and Nebraska revealed many similarities of purchases among ethnic, cultural, historic, environmental, and recreational travel types. However, special items were found to relate to each

type, for example, wooden crafts, postcards, and photographs for outdoor recreational tourists. A study of communities in central Alberta, Canada (Getz, Joncas, and Kelly 1994), revealed the significance of the travel segment of shoppers in that location. Japanese travelers engage heavily in shopping. They and other travelers seek villages with a small town atmosphere, where they feel welcome, and where convenient amenities are provided. Shopping demand is often linked to dining.

A significant cultural travel segment involves visiting places where prominent authors and artists lived and worked: literary tourism. Squire (1994) has documented the great volume of visitors to the area in England where Beatrix Potter wrote *The Tale of Peter Rabbit, Flopsy Bunnies, Johnny Town-Mouse,* and many other writings. This segment of travel is often linked to heritage and genealogical interests.

TRENDS

An important way of viewing travel markets is to recognize their dynamics. Markets change; therefore, every year developers of tourism must keep up to date on trends. For example, Dychtwald (1989) made an analysis of changing American markets, concluding that: middle-age physical, psychological, and social concerns would become more important; nontraditional lifestyles and distrust of authority would continue; middle-aged people would strive for a balanced lifestyle; health and quality factors would dominate destination choices; the importance of convenience and comfort would increase; and experience rather than things would become more important.

As of this writing, there are several trends worth considering by investors, planners, developers, and community leaders (Gunn 1995).

From Consumption to Nonconsumption

Hunters are photographing wildlife more than killing it; the experience of being outdoors and appreciating natural resources is becoming of greater importance than harvesting game. Sports fishermen are increasingly

TABLE 3-2. Cultural Activities of Visitors to the United States

Activity	UK	France	Germany	Japan
Concerts/plays	31.9	38.3	24.0	9.3
Local festivals	39.2	35.3	38.9	19.3
Museums/galleries	37.1	63.1	32.1	13.9
Historic places	52.6	56.5	54.1	17.0
Commemorative places	37.5	57.3	42.9	5.9
Archaeological places	17.7	19.6	25.7	3.1
Military historic sites	30.7	21.2	20.3	8.7

Source: Tighe 1990, 14.

releasing their catches; the fun is seen to be in the total experience on waters and in making the catch. Although trophy hunting and fishing continue, the trend is toward nonconsumptive activities. More travelers are interested in protecting rather than destroying resources. Some of the most vocal environmentalists today are travelers who see increasing problems when mass tourism disturbs and even destroys natural and cultural resources.

From Commonplace to Sophisticated

More travelers are seeking greater depth of experiences than in the past. The ordinary no longer satisfies. Travelers want to understand how things work and how they came to be. The mundane and shoddy, once popular with less sophisticated travelers, is no longer acceptable. Travelers are more experienced and more highly educated. Developers must respond with greater opportunities for guidance and interpretation. Tour guides, for example, must abandon rote narratives and be knowledgeable in depth about the sights the visitors are experiencing.

From Younger to Older

Worldwide, demographics are changing, especially toward longer life. Most older travelers are very experienced and demand better quality. However, this group must not be considered homogeneous; they are as diverse as they were at younger ages. Many are still physically able and wish to participate in challenging physical activities. Others prefer more passive experiences and opt for motorcoach tours and spectator entertainment. Many in their sixties and seventies have ample funds for travel and seek high-level accommodations, food, and attractions. Supply-side developers must be alert to this new travel segment and respond with services and activities suited to a variety of needs.

From "I, Me" to Greater Interest in Others

Although individual rewards from the travel experience remain important, there is a trend toward greater interest in other peoples. Perhaps because of more education, more telecommunications, or more travel experience, many travelers today seek to know and understand other peoples of the world. News about wars, unrest, and political changes is widespread, providing new insight into areas less well known in the past. This trend has significant impact on the development of the supply side of tourism.

From WASP to Color-Blind

There was a time when travel was dominated by White, Anglo-Saxon, Protestant members of society. Today,

there is an increasing trend toward greater diversity of travelers in nationality, race, color, and creed. Many destinations are providing greater opportunity for travelers of many categories to visit their areas. Ethnic, racial, and national groups that heretofore did not travel are now sufficiently affluent to do so. Many areas of the world are not yet prepared to respond to this travel demand.

From Domestic to International

Rapid growth in the economies of many newly developed countries and the greatly expanded air service around the world are offering travel opportunities to millions more international travelers than in the past. These travelers of different traditions, languages, customs, and interests provide new challenges to the areas receiving them. They are being courted heavily today because generally they spend more money at their destinations. Supply-side developers cannot simply use the methods and provide the services and attractions that were suited to domestic travelers of the past if they want to attract the travelers in this segment of the market.

It must be emphasized that these are current trends that will most likely change in the future. All travel markets are dynamic; they are highly dependent upon changes in policies, interests, and economic opportunity in the countries of origin. This fact highlights the need for developing areas to gain access to the latest travel market trends and perhaps initiate their own market research.

From Business-Only to Business-Pleasure Mix–Mix

Established several years ago and continuing today is the trend for travelers to combine business and pleasure in their trips. Spouses and families often accompany the business traveler on both domestic and international trips. Convention and conference managers have discovered the popularity of pre- and postconvention tours that provide travelers with rich scenic and cultural experiences in and around the meeting venue. Many corporations continue to reward their most productive employees with incentive travel: If a certain business quota is met, the employees and their families are rewarded with an international trip. Industrial leaders, as they visit potential corporate expansion sites to evaluate the local amenities and suitability for future living, are often accompanied by their families. Even single business travelers combine business and pleasure activities on many of their trips.

From Generic to Specialized

The trend toward greater specialization among travelers is increasing (Hollinshead 1993), reflected in the increase in numbers of travelers interested in such purposes as adventure, ethnic experiences, culture, history, archaeology, bird watching, diving, and encounters with local citizens. For this group, each trip tends to satisfy primarily a single interest, although occasionally there is overlap. One trend among specialized travelers is the tendency to select other than traditional accommodations. A study of U.S. tourists (Morrison, Pearce, Moscardo, Nadkarni, and O'Leary 1996) revealed a significant market segment that prefers small lodging establishments where close contact with owner-hosts is available and where an atmosphere of vernacular architecture can be obtained. This trend is exemplified by an increased popularity of bed-and-breakfast accommodations in restored older homes. An understanding of this trend toward specialization is helpful for communities seeking tourism development.

TRAVELER IMAGES

It is important for designers and developers of tourism to recognize the psychology of traveler images, from the beginning to the end of trips. Several decades ago, psychologist Bruner (1951) described a three-phase process that may be applied to travelers. As illustrated in Figure 3-2, the sequence consists of: hypothesis, input, and check.

Travelers tend to bring along images of destinations that they have accumulated over their lifetimes. Even before they reach a destination, they visualize themselves in that setting. *Hypothesis,* or expectancy, explains why different visitors may have quite different reactions to the same stimulus. For example, a person familiar with the habits of birds may spot an eagle's nest more than 300 yards away, whereas another individual, perhaps with greater interest in historic buildings, may not see it at all. Expectancy gains strength in several ways. The more the expectancy of travel experience has been confirmed, the more confident the user is of potential satisfaction upon repeating the experience. Tourists are likely to see what they anticipate seeing.

The next phase takes place when the traveler actually experiences the intended objective. This step is the critical point in both touring-circuit and longer-stay activities when the material truth is revealed. The *input,* or stimu-

lus of a place, incites reaction through all senses—vision, hearing, pressure and touch, temperature, kinesthesis, pain, taste, smell, vestibular sense, and common chemical sense. Often, several senses are intertwined. For example, one experiences a mountain climb through the lungs and by muscles as much as through the eyes and ears. A waterfront sunset may exert as much impression through the sound of the waves, the smell of the water, the sounds from shore birds, and feel of the cool breeze as from the visual impact of the many-colored setting sun. The totality of the experience leaves enduring impressions.

The final phase is a comparison between what was hypothesized and the real input of the experience, which may be similar, better, or worse than the anticipation. This *check* is a test of congruency with the traveler's original attitudes and images. A heavy responsibility falls on the roles of designers and developers to create attractions that, ideally, will exceed all expectations. Unfortunately, the realm of hypothesis is not within the control of developers; it is the result of the traveler's lifetime of mental accumulation from a great many sources. Of some help to developers is the information resulting from tourist behavioral research. Such studies can offer clues to the images people have at their homes of their travel destinations.

Communities that anticipate tourism growth can guide their development best when they have information on what would-be travelers believe about them.

CONCLUSIONS

Although tourism demand and supply can be described as two elements, they are really only two sides of the same thing—the tourist experience. What people want as they travel is largely derived from development that is available. Conversely, what is developed is a reflection of what travelers want.

In the process of development, especially at the local level, the demand side must be understood. For any des-

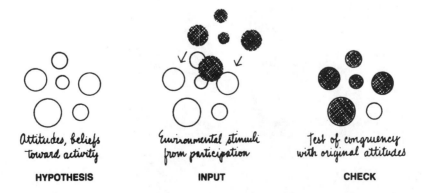

Figure 3-2. *Diagram of travel image psychology. Three steps in the traveler's experience test the "fit" between what was anticipated and the reality of participation—an important principle of development.*

tination, key factors concerning markets include origin, socioeconomics, policy constraints, motivation, interests, and activities. No longer can one generalize about traveler characteristics. The great diversity of travelers is sometimes divided into market segments as useful concepts for planning. For example, the category of ecotourist has emerged recently and is already a significant segment that promises great growth.

It is especially important to understand that the travel market demand side is very dynamic. This characteristic militates vigilance among tourism planners in reviewing trends. With the knowledge such vigilance provides, planners can place emphasis on developing supply and on measuring how well the development can not only satisfy existing demand but also be sufficiently creative to stimulate new demand.

As travel demand continues to grow, planners and developers are obligated to solve issues of capacity. Every resource must be evaluated for determining levels at which the resource and the visitor experience will deteriorate, such as on the Au Sable River, Michigan.

Especially vulnerable to increased visitor demand are cultural sites, such as this Castle of Chillon, Switzerland, immortalized in Byron's poem, "The Prisoner of Chillon." Increased water sports and cruises as well as masses of visitors inside can erode the resource and diminish the value of the historical visit. (Photo courtesy Swiss National Tourist Office)

CHAPTER 4

Tourism Function: Supply

Most tourism supply development is sparked by individual decisions—a new hotel, a new attraction, a new festival. But these are only pieces of the whole; much more is involved in the functioning of tourism. Perhaps this principle is best understood by the metaphor of an automobile. Essential parts include the battery, wheels, power train, and engine, today critically influenced by computer controls. Malfunction or failure of any one of these parts seriously impairs the car's overall performance. So it is with tourism. It cannot function properly unless each part is doing its work most efficiently and in close harmony with all other parts.

When communities seek to expand or newly develop tourism, they must not only avoid the negative impacts but also make sure that investment in new parts provides the best overall functioning of tourism. For example, there is no need to develop an attraction in a remote location where access is difficult or impossible. There is no need to build a new hotel unless new attractions are developed to bring travelers to that destination. There is no need to spend great amounts of money on promotion unless there are viable attractions to promote. And there is no point in making large investments in development if they do not meet market wants and needs.

The supply side of tourism can be defined as all the physical development and programs that provide for the needs and desires of travelers. It is clear from this statement that a great many developers in a diversity of components are involved. It is equally clear that all parts of the supply side are directly related to the characteristics of demand.

THE FUNCTIONAL TOURISM SYSTEM

Figure 4-1 illustrates a model of the *functional tourism system*. First, it shows the intimate relationship between the demand and supply sides of tourism. Each is dynamic and has impact on the other. Second, this model defines the supply side of tourism as composed of five compo-

nents: *attractions, services, transportation, information,* and *promotion*.

When a community, state, province, or country contemplates improvement or expansion of tourism development, it has to consider the supply side in its totality, not just a few parts. Even though successful operation of each part is essential, equally important is how the many parts interrelate. The problems of integration are exacerbated by the marked differences between tourism and other economic development.

For one thing, tourism distributes markets to products (travel destinations) rather than distributing goods to markets from points of manufacture. This difference means that the product areas, the places to which those markets travel, are more difficult to plan, design, and

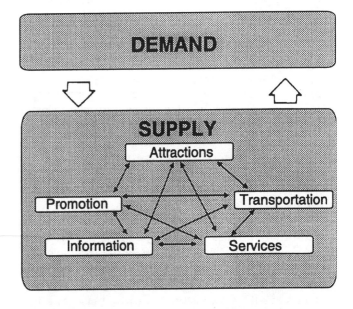

Figure 4-1. *The functional tourism system. The supply side of tourism (development) should be in balance with demand (markets). Supply is modeled to include five interdependent components: attractions, transportation, services, information, and promotion.*

manage. For the traveler, these places mean everything, whereas the place of manufacture of goods is of no concern to the consumer. The tourism product is not canned or shrink-wrapped but, rather, a collection of experiences gained by the traveler.

Tourism supply, therefore, is indigenous; it is of and by the land. The natural and built environment for tourism varies throughout a nation. Some areas have more settings of interest to travelers than others. The location of a manufacturing plant is less dependent on such settings.

Because tourism development at the community, regional, state, or national scale includes thousands of properties and their owners, it is difficult to realize how they may be interrelated. If problems of poor connectivity and integration are to be solved, a broad perspective of how tourism functions is critical. Tourist organizations and agencies at the larger scale can be instrumental in increasing coordination and integration of tourism if they plan for its functional entirety rather than merely its separate parts.

One way of gaining insight into how tourism functions is through the scenario of a trip from home to destination and back. The travel itinerary depends on many informational and promotional factors. For example, the anecdotes of a friend, a magazine article, an advertisement for a vacation package, or a book or movie may influence the selection of a destination. Personal preference may influence the choice between a touring-circuit or resort-type vacation. Income and personal preference may influence the mode of travel—tent camping, recreational vehicle (RV), caravan, historical tour, air, cruise, or personal car. If an air travel or motorcoach package is selected, a travel agent probably books the entire vacation, or portions of it, such as air travel and lodging.

On the way, the auto traveler makes many decisions on places for food, lodging, and car service. The need may arise for banking, telephone, or health service facilities en route. Throughout the trip, the traveler relies on maps, signs, and guidance from local residents for directions. Based on planning before leaving home or impulse along the way, stops at attractions may be added to the touring circuit. The RV traveler is likely to visit natural resource sites en route to a destination, enjoying the flora and fauna of parks and recreation areas.

The experiences obtained depend on a successful match between the supply development (especially attractions) and the traveler's preference. For example, if a community has restored many of its historic buildings, developed an outdoor drama of its history, and replicated period crafts for sale, history buffs may make sure they visit there. Public agencies that manage extensive forests, lakes, and wildlife resources may attract many visitors to view, photograph, or become enriched by these environments, even though resource protection is their main objective.

Other travel segments may choose urban destinations because of their special attractions. Historic sites, entertainment halls, specialty food places, museums, theme parks, sports arenas, and the homes of friends and relatives attract visitors seeking the particular experiences they afford. Of course, convention centers, trade centers, and a variety of business and professional offices make cities the objectives of business travelers.

Upon returning home, the traveler reflects on the experience. Discussion with friends and relatives reveals the extent to which the trip has fulfilled expectations. The greatest adventure may have come from difficulties encountered along the way. Through snapshots, color slides, videos, and souvenirs, the traveler relives the trip and relates it to others. The success or failure of the experience may have a great influence on next year's trip plans.

As can be seen by this description, there are a great many travel scenarios, because the various segments of the population have different preferences for travel experiences. Compared with expenditures of time and money on other activities and goods, travel must have high priority. Furthermore, the traveler must have an income sufficient to fund it.

The point of this discussion is to demonstrate that overall travel involves many business establishments, many public places, and many programs by organizations and agencies directed toward specific functions. This complexity of places and programs functions well only when each individual part is designed and managed so that it relates well to the others, in addition to carrying out its individual, primary function. It is not enough, for example, that a hotel has good facilities and services and is well managed. Unless it is accessible and related to the surrounding attractions that bring visitors, it is not carrying out its complete function. It is a lack of interrelation between the many parts that has often caused travel difficulties. Although the planner, developer, or community leader cannot solve all such problems, much improvement can be accomplished when all decision makers understand the major components of the supply side and the interrelationships among the many parts that bind tourism together in a system.

Transportation

Although the improvement of technology for transportation modes in the last few decades has greatly improved many aspects of travel, many consumer difficulties remain. The manyfold increase in the speed of jet planes over that of earlier propeller-driven planes has stimulated travel to destinations never before accessible. Shorter periods of time in the air and less upset by turbulence have increased traveler comfort. Improved automobile design and construction have produced more efficient and comfortable cars. The divided freeway concept has

The Sabi Sabi safari resort of South Africa demonstrates the functioning supply side of tourism. It is accessible by flying and driving from Johannesburg or Pretoria; it is located near the wild game attractions of both private game reserves and Kruger National Park; it is served by the communities of Sabie, Skukuza, and Nelspruit; and it provides guided photography tours and deluxe lodging and food service.

Wildlife photographic and sightseeing safaris give the visitor enriching opportunity to observe more than 200 species of animal life, including lion, rhinoceros, buffalo, leopard, elephant, cheetah, zebra, monkey, giraffe, and wild dog. The design and management challenge is to control mass tourist use so as to maintain environmental assets.

Even though modes of passenger transportation have improved greatly since the era of Ford Model Ts and the muddy roads of the 1920s, there remain many issues of improved transportation before tourism planners and developers. (Photo courtesy Texas State Department of Highways and Public Transportation)

allowed much greater volumes of traffic and at the same time has reduced the rate of accidents. Tourism would not have become such a dominant worldwide phenomenon had it not been for these improvements. Even so, more changes are needed. It seems that there is much travail in today's travel, just as described by Jerome Turler in the sixteenth century: "Nothing else but a painstaking to see and search for foreine landes, not to be taken in hande by all sorts of persons or inadvisedly, but as are meete thereto" (Turler 1951, 5).

The miracle of jet travel is often offset by difficulties with which only the calloused veteran can cope: missed connections, overbooking, pilot errors, mechanical problems, controller errors, chaotic terminals, and even hijacking. The miracle of freeway travel is offset by poor directional signage, drunk and otherwise impaired drivers, truck driver bullying, and the violence of some frustrated

motorists. Within cities, the application of traffic engineering has improved traffic flow at the expense of livability and, especially, pedestrian use of downtown areas. Many years ago it was learned that the economic benefits of tourism are not derived from people in motion; visitors must have the opportunity to stop, leave their vehicles, and enjoy the amenities on foot.

Those who own, design, and operate transportation systems used by tourists have a narrow view of users because of the limitations of their agencies. Highway planners, for example, usually improve highways only when accident rates increase, rather than when new developments create new demands. Highway officials usually see travelers as part of a mechanical flow diagram, with each mile of concrete ribbon calculated on the basis of so many rubber-tired steel units per hour. Airline officials generally regard travelers as point-to-point cargo, to be made airborne at one place and then grounded at another at the fastest possible speed. Airport decision makers are mainly concerned with servicing aircraft; passenger ticketing, toileting, and feeding; and lost baggage processing. And all of these tasks are accomplished within heavily invested capital improvement programs that are only incidentally related to attractions, other transportation modes, and the regional tourism environment. In most countries there is no coordinating agency that relates transportation policy with tourism (Cunningham 1994).

Recently, transportation planners have increasingly studied and developed concepts for improved movement of people, particularly within cities. Although no simple solutions have emerged, communities need to avail themselves of all innovations that may be useful, especially for visitors. Although traffic engineers have been enhancing the mass vehicular movement of people, the most important form of transportation—pedestrianism—has been neglected.

As tourism interest in older and restored city cores increases, such as at the French Quarter, New Orleans, the concept of traffic-free areas has expanded. Either by the complete banning of cars and buses or by time-zoning, these areas have become exciting, safe, and worthwhile places for locals as well as visitors. Although merchants often resist such innovations, evidence now suggests that conversion to pedestrian zones increases sales in most parts of the world (Lennard and Lennard 1995).

Lennard and Lennard (1995, 77) have urged that the principle of balanced transportation be applied to all communities, according to the following rules:

1. Accommodate the real needs of people;
2. emphasize access, not traffic;
3. balance transportation with land use;
4. use care in mathematical modeling;
5. set up a hierarchy of modes;
6. consider social functions;
7. use parking restrictions;
8. design to human scale;
9. maintain people-use of ground level; and
10. improve livability and aesthetics.

Properly designed pedestrian spaces enable people to move in all directions without conflict. Walking encourages meeting people, exchanging greetings, and obtaining guidance—especially important for visitors. Walkway amenities, such as park settings, seating, shelters, and snack bars, add much to the social value of these places. In natural areas such as national parks and cultural areas such as historic buildings, the social exchange among visitors is as important as contacts with nature or history.

As attraction complexes get larger, the need for slow-speed people-movers may arise, as demonstrated in major theme parks. Historic sites often employ vintage horse-drawn carriages now that sanitation techniques have improved. Minibuses, monorails, tractor-trailers, excursion boats, funiculars, hydrofoils, and narrow-gauge rail systems provide fun as well as efficient movement for people.

Way-finding, an essential part of transportation, is often neglected. Scattered, illegible, ambiguous, or too plentiful signs, together with cluttered landscapes, bear testimony to the low design and management priority of visitor orientation. The lack of integration between sites results in an environment that not only is frequently unsightly but often fails to help travelers find their way. One can test this truth simply by posing as a traveler and taking note of every instance of inadequate directions. The results will quickly convince anyone that a problem exists. Romedi Passini (1984, 165) observed the same problem for residents: "In our everyday environment, it is not complexity in itself that creates wayfinding difficulties but the combination with inadequate design, leading to featureless settings or conditions of overload." The need is great for better design and management.

The work of Passini, as well as of environmental behaviorists and landscape architects, emphasizes the needs for an improved understanding of the individual's relation to his environment and for better design of clues and cues. Passini has described way-finding as a cognitive process requiring: a cognitive mapping ability to allow the traveler to gain an understanding of the environment; a planning-for-action ability; and a decision-making ability to transform plans into action.

Directional signs (arrows, distance markers) help travelers make choices. Yet they are often ambiguous, or poorly designed, or placed so that the message is never transmitted. Identification signs need to tell the traveler that he has arrived. The place, object, or person must be clearly stated, frequently by affixing the sign to the destination or building. Many travelers will attest to the need for reassurance signs; after changing directions or traveling many miles and meeting many intersections or route choices, it is comforting to know that one is still on the right road. An art in itself, especially for interpretation at sites, buildings, and museums, is the design of informative signs. Too often these are crammed with words printed in too-small type. Again, the time and interest span of the visitor offers the best clue to the quality and quantity of information that should be included.

Landscape design itself can offer clues for way-finding. For pedestrian areas, paving materials of different colors and textures can effectively lead the visitor in the direction intended. Color coding, uniform styles, and coordinated shapes are effective sign design techniques. Well-placed kiosks where notices can be posted are useful but require maintenance and control. New materials allow weatherproof displays of maps. But engineers' maps need to be redesigned with simplified diagrams for greater readability by visitors.

Services and Facilities

Individual building and landscape designs for tourist services have gradually improved in aesthetics and efficiency, in part because of increased public demand for better quality. Franchises and corporate developments, as compared with "mom and pop" establishments, can now afford sophisticated market surveys and management practices. For fast food chains, these techniques have resulted in attractive designs of interiors and exteriors, better building styles, and better landscape development. Some motel and restaurant chains, yielding to local demands for the adaptation of commercial establishments to particular environments such as historic areas, have more than one design for sites, buildings, and signs.

Even so, individual decision making generally remains nearsighted rather than broad-visioned, and the free enterprise system is not responsible for this myopic view. Rather, a lack of understanding of the true nature of each business purpose and a misconception of the overall supply product are at fault.

Great changes in food production, especially the United States, are taking place, such as the "cook-chill" process. This technology involves cooking meals prior to service, then rapidly chilling them, refrigerating them until needed, and finally rethermalizing (often by microwave) and serving them (Green 1993). This process allows food preparation at some distance from retail sales and distribution while it is refrigerated. It claims lower food costs, greater consistency of quality, lower labor cost, and increased health and safety. Restaurant location trends are also changing. McCool (1993) has cited three current types: isolated, within other businesses, and adjacent to related food or other business.

The design and development of lodging remains narrowly site-oriented. To satisfy the several market segments, the new generation of lodging ranges from bed-and-breakfast and budget motels to luxury hotels with large suites and hot tubs. However, owners and managers of lodging continue to be preoccupied with operations within the site at the expense of understanding the many external influences on their success. Room sales are as dependent on access, relation to attractions, and neighborhood amenities as on the quality of the hotel. Unfortunately, neither educational institutions nor trade associations emphasize these external factors.

A major innovation in lodging management in recent years has been the multinational firm (Go 1993). Such transnational firms have resulted from the emergence of new business centers, economic growth of travel origins, government policy (incentives as well as barriers), and international travel demand. Further stimulation has come from economies of scale, greater research ability, and ready access to new technology. However, there is increasing concern over some negative impacts of this trend—excessive standardization; insensitivity to local social, economic, and land use issues; and lack of flexibility for changing travel markets. McNulty and Wafer (1990) have called attention to the greatest vulnerability of resource-poor nations to transnational impact. Massive enclave tourism, such as is associated with casinos and high-rise hotels, can cause severe social clash with local residents. Mass tourism established by transnational firms can be destructive to cultural customs and artifacts. McNulty and Wafer have recommended the following measures to reduce these negative impacts: prohibiting foreign land ownership, emphasizing local materials in construction, restricting payments to foreign management, and guaranteeing some local investment by limiting repatriation of profits.

Alongside multinational trends has developed the concept of franchising, an American original that is slowly becoming international. Lavin and Lunceford (1993) have cited three trends in franchising: consolidation and mergers of firms, catering to new market segments, and worldwide acceptance.

Commercial services for travelers greatly overlap those for residents. Restaurants, car service stations, and shops are used both by resident and travel markets. Even hotels, catering primarily to travelers, frequently have resident services such as restaurants, gift shops, and banquet rooms. However, urban politicians, planners, and designers tend to consider only resident functions of commercial services when they make policy and planning decisions.

Here and there across the land, as illustrated later in this book, community leaders, designers, and owners of services for travelers are becoming more sensitive to the need for better, more relevant design. Exterior and interior designs of hotels, restaurants, and shops are beginning to reflect regional themes—verdant, desert, ethnic, or historic. For example, old decorative iron facades along the Strand in Galveston, Texas, (once a warehouse district,) not only carry out the area's 1890s theme but also lead visitors to modern food services and tourist gift shops inside.

Travel agents and wholesalers are undergoing a communication revolution because of new computer programs, but personal contact remains an essential part of service. It is increasingly important for service managers to be aware of trends and understand the many complexities of their roles. As an intermediary between the prospective traveler and all the supply-side components—accommodation, transportation, attractions, and the management mechanisms for reservations—the travel agent has an ever-increasing burden of factual information (Wohlmuth 1994). The impact of the threat to these functions from the personal computer and direct airline reservation and ticketing services is not yet known.

The services component of the tourism system deserves greater local community attention for a better relationship between tourist sites and surrounding visitor functions and landscapes.

Information

State, national, and community tourism agencies usually carry on programs of information and promotion as if they were the same. However, more careful scrutiny of these functions shows their difference.

There are several reasons for increased concern over the topic of travel information today. The explosion in the number of places to visit, because of both easier access and the greater propensity of people to travel, has increased confusion on the part of the potential traveler. There are so many choices and so many promotional pieces and advertisements that travel decisions are harder to make. The growth of travel agent services, guidebooks, computer travel programs, popular magazines, and guided motorcoach tours bears testimony to the acceptance of new and better travel information.

Because most travel market segments are more sophisticated today—in terms of both education and travel experience—they are seeking more and better information. They are no longer satisfied with merely being exposed to an attraction. If they are visiting a historic site, they want more description of its relevance to events of the period. If they are visiting an outdoor recreation area, they want more information on how to hunt, fish, or photograph wildlife. Because the population has become more urbanized, most travelers no longer have the close contact with nature common to previous generations, and more things need explanation.

Tourism developers now have an important role in fostering better information provision for visitors through greater use of interpretation centers. In the fields of outdoor recreation and parks, the term *interpretation* generally applies to the process of helping tourists understand and enjoy the environment they are visiting (Machlis 1986). Communities are creating new visitor centers and ancillary services, such as trailside information, to handle mass tourism at park sites. Included often within the centers are exhibits, dioramas, demonstrations, and lectures. Often trails are designed to bring the visitor into close proximity to special features of the landscape. Along the way, numbered stops, keyed to explanations in an information booklet, help the visitor understand the environment. When properly designed, such interpretive areas can provide a great number of visitors with satisfying experiences without disturbing and eroding resources (Gunn 1994). In other words, a high percentage of travel markets are satisfied with a somewhat vicarious experience that does not demand close contact with or disturbance of natural and cultural resources.

A continuing controversy is over the use of roadside signs. Because the experience of viewing roadside scenery continues to rate high in travel surveys, it seems incumbent upon designers of highway corridors to keep them attractive. Generally, travelers prefer natural resource vistas, farm landscapes, and well maintained structures over excessive billboards, junk yards, and dilapidated buildings. A private U.S. organization, Scenic America, advocates improved roadside scenery, especially the designation of scenic roads. Approximately $1 billion in tourist revenue is generated annually from scenic highway programs (Scenic America 1991). Essential to scenery is billboard control. In recent years, many local and state governments have enacted regulations for such control. Examples include a complete ban on billboards in Vermont; a Raleigh, North Carolina, ordinance in 1983 that requires removal of billboards in certain areas (by 1992, 100 had been removed); and more than 50 destinations (cities and states) that have prohibited billboards.

On the other hand, travelers do seek directions as they travel. They need enough highway signs to direct their course, and they need some means of identifying the locations of services and attractions. A collaborative study by designers, planners, highway engineers, and local government agencies responsible for roadside zoning is necessary to determine how best to fulfill these needs. Today, electronic location and guidance technology is making new installations in automobiles available for traveler information.

Promotion

The component of promotion encompasses all forms of enticement and persuasion used to alert the market to travel. From the perspective of travelers, however, the limits of the power of overt enticement must be recognized. The extent to which market populations understand destinations and their features can be described in two ways—organic and induced.

By the *organic* image of a destination is meant the totality of what a person already knows or perceives about that destination. A report from Alaska, for example, stated, "Most of our vacationers have been assimilating impressions and information about Alaska for a number of years" (Hinkson 1964, 6). Images are accumulated over time from newspapers, radio and TV news, documentaries, periodicals, dramas, novels, and nonfiction. Books and classes on geography and history give children insight into the natural and cultural resources of areas. Word-of-mouth narratives from friends and relatives returning from trips can be major influences on people's impressions of destinations. All these influences combine to give individuals, right or wrong, their impressions of the characteristics of travel destinations and how rewarding or disappointing they might be. Although it has not been measured, it is likely that this organic accumulation of information is the most powerful factor influencing travel decisions.

More familiar to tourism agencies and businesses is the

Too often, roadside environments are visually littered with massive and confusing signage. This practice, founded in the desire to improve business, actually is questionable promotion and blocks scenic vistas that are important to travelers.

realm of *induced* images of destinations—overt processes designed to attract travelers to certain target areas. There are four popular forms of induced image development. *Paid advertising* consumes millions of dollars' worth of artists', writers', and photographers' work. This money is lavished on the print media (folders, newspapers, magazines), radio, and television. *Publicity* is often used by tourist organizations. Performers and craftspeople are sent to market areas to publicize the special features of a destination. Participation in travel shows allows organizations to display exhibits, folders, and videos of destinations. *Public relations* is a more subtle method, whereby representatives of destinations appear at meetings and conferences on tourism and related topics. *Incentives* are quite popular with tourist businesses—in the forms of coupons, discounts, packaging, and contests.

Attempts to measure the effectiveness of these promotional methods are called *conversion studies.* Researchers and scholars continue to debate the effectiveness of promotion in bringing tourists to the intended areas. Some studies have indicated that would-be travelers are far more influenced by the many organic factors. Destination marketing organizations recommend before- and after-trip surveys to determine awareness, image, and effectiveness of advertising campaigns (Perdue and Pitegoff 1994).

Attractions

Without developed attractions, tourism as we now know it could not exist; there would be little need for transportation, promotion, facilities, services, and information systems. The principle of attraction is so important

that it deserves deeper examination by community leaders, developers, and planners, especially as the growth and expansion of present environments are considered (see Chapter 5).

SUPPLY DYNAMICS

The supply side of tourism represents the true tourism product. All developers must view this product from the visitor's perspective. Smith (1994) came to the conclusion that the tourism product has five dimensions based on visitor use and management control: The *core* is the physical plant—land, water, buildings, equipment, and infrastructure. The first ring around this core is *service*—those tasks needed to provide for management, operation, and maintenance. The next concentric ring is labeled *hospitality.* Whereas service refers to the technical operation of tourism, hospitality means hosting attitude and performance. *Freedom of choice* refers to the visitor's opportunity to choose activities and places to visit. Spontaneity is as important as the variety of choices. Finally, farthest from the core is *involvement* by the tourist. Actual involvement is the measure of satisfaction that compels visitor participation in the experience. Smith emphasized that as one progresses outward from the core, management control decreases while visitor participation increases. This view of the supply side stresses the relationship between visitor experience and the physical and program aspects of development.

Once tourism leaders and supporters in a community understand the five components of supply and how they relate to the market, the need for guiding all parts into a functional system becomes clear (Figure 4-1). It is insufficient to expand promotional programs without understanding the status of attractions, services, transportation, and information systems. It is a mistake for a community to accept a major resort development without a thorough understanding of how it relates to the travel markets and the potential impact on the community's societal, environmental, and economic status. A new motel or hotel should not be encouraged unless there is clear evidence of traveler demand for new lodging. Any change in any part of any component by definition influences the operation and success of all other parts of the system. The more a community strives for balance among all the components of supply, and especially with market demand, the more successful its tourism will become.

SECTORS

The most popularly known sector of tourism development is the commercial enterprise sector. This sector is dominant in market economy countries and includes hotels, restaurants, and entertainment. However, throughout the world two other sectors are equally important as tourism developers—the public sector and nonprofit (volunteer) organizations.

Commercial Enterprise Sector

The concept of commercial enterprise was first publicized by Adam Smith (1930) in 1776, in his *The Wealth of Nations*. The concept, as Friedman and Friedman (1980) have pointed out, was deceptively simple: If an exchange between two parties is voluntary, it will not take place unless both believe they will benefit from it. Applied to tourism, this principle simply means that travelers are willing to trade some of their earnings for products and services they seek. The controlling mechanism that decides the trade is price. If the quantity and quality of service or product represent satisfaction at the given price, the trade will take place. However, if quality and quantity are not satisfactory, the exchange will not take place. If this relationship is tampered with—by what is sometimes called *static*—such as through government intervention, the system breaks down. Among the many messages sent by the relationship between buyer and seller is the clue for producers to offer what the market wants. In a free market situation, for example, a hotelier soon learns what type of rooms to offer and at what price. If, however, the operation is subsidized by government, this message is no longer clear.

Allen, Armstrong, and Wolken (1979) produced a concise summary of how this system works in *The Foundations of Free Enterprise*. The basic tenets of free enterprise, according to these authors, are private property, economic freedom, economic incentives, competitive markets, and limited role of government.

Private property

Certain property rights are essential: the owner's right to determine how his property is used, the owner's right to transfer ownership to someone else, and the owner's right to enjoy income and other benefits that come his way as a result of his ownership. Private property tends to be abused less than public property because it is personal wealth.

Economic freedom

Business owners enjoy several privileges: the right to start or discontinue business, the right to purchase any resource they can pay for, the right to use any technology, the right to produce any product and to offer it for sale at any price, and the right to invest in any way. Such stimulation of self-interest encourages production, which in turn leads to higher standards of living.

Economic incentives

The desire to improve stimulates greater productivity and efficiency, which, in turn, usually produces greater rewards. The distribution of these rewards answers the question of what to produce or serve. Scarce resources, labor, and creativity are directed toward the products and services valued by other members of society. Equally important are the punishments when these efforts fail to produce what society needs and wants.

Competitive markets

The force of competition provides a large number of buyers and sellers rather than monopoly control. Producers are free to compete for the consumer's money. Each individual consumer makes his vote for a product or service when he decides to buy or not to buy. The great diversity among tastes and preferences of consumers is far better than a bureaucratic decision that eliminates competitive markets.

Limited role of government

Free enterprise is founded on the principle that individuals, not governments, know best how to pursue their own well-being. Under this theory, government performs its role best when it allows individuals to engage in free enterprise. Other than creating rules for society as a whole and enforcing these rules, government has no economic responsibility. As regards tourism, governments may be involved in land development and management through their social welfare functions (see below).

These five characteristics are the foundation for the best function of free enterprise within a market economy.

Of course, there is no pure free enterprise activity; it is tempered by actors' honest desire to improve society, often with accompanying government intervention. But doing good with other people's money has two flaws. One never spends other people's money as carefully as one's own. And spending other people's money means getting it first—not an easy task.

A major impetus for government regulation is the desire to control business that is unscrupulous, unfair, or even fraudulent. Within tourism today there is a major countermovement, for business to control itself. Businesses are beginning to accept their responsibility for environmental protection, enhanced societal and individual values, and reduction of economic burden. (These ethical issues are further discussed in Chapter 2.)

In regions such as undeveloped countries that are without a strong tradition of entrepreneurship, new training and education may be needed. Rather than depending on outside multinational investments, these regions can achieve a better integrated tourism in time by giving local people the education on free enterprise that will enable their own investment in tourism.

The leaders of the province of Nova Scotia in Canada sensed the need for a greater understanding of the role of entrepreneurship in tourism and sponsored the examination of many case studies (Acadia Institute of Case Studies 1995). The purpose was to identify exemplary establishments that had demonstrated high-level management skills. The resulting publication includes case studies of businesses such as Bev's Ceramics, Lower Sackville; Cambridge Suites, Halifax; Green Acres Farmers Market, St. Mary's; Linda's Bed & Breakfast, Westside; Mountain Gap Resort, Smith's Cove; Peddler's Pub and JJ Rossy's Ltd., Halifax; and Wool'Uns Quality Woolcrafts & Garments, Pictou. Accompanying this study is an instructor's manual for use in colleges and universities.

In many undeveloped and developing nations, land tenure is often a touchy issue. If complete title to land ownership is not available, potential investors are reluctant to establish capital investment in development. Government policies need to be updated and clarified so that fee-simple ownership of land is a legal fact.

Research by Snepenger, Johnson, and Rasker (1995) has revealed another important facet of entrepreneurship—"travel-stimulated entrepreneurial migration." Study of businesses in the area around Yellowstone National Park showed that four out of ten of today's business owners first experienced the area as visitors. Attracting visitors can stimulate the migration of potential investors and developers.

Although profit making is an incentive for business, profits are not merely personal wealth for the owners. Out of profits must come insurance, property taxes, employee benefits, replacement and repair of equipment, and adjustments to changes in market preferences. These are socially and economically sound attributes of the commercial enterprise sector. Because this sector depends greatly on external attractions for its success, it owes some of its profits to those attractions for their support. Certainly, the commercial enterprise sector is a major player in the development and continuance of tourism.

The modern Outback Steakhouse chain of more than 300 locations in the United States illustrates a successful application of Allen's principle of free enterprise (Sullivan 1996). A management-ownership program provides not only an economic incentive for high-quality product and service as part-owner, but also assures greater than average personal income. This innovative approach allows the owner-manager to take part in community activities and voluntary organizations as well as to efficiently adapt food service to market needs, which is generally prohibited by other chains.

Nonprofit Volunteer Organizations

Worldwide, a second sector, that of nonprofit organizations (and volunteers), is as important for tourism development as the commercial enterprise sector. A great number of physical development projects and programs for tourism are sponsored by health, religious, recreation, historic, ethnic, professional, archaeological, and youth organizations. Because their objectives are oriented more toward culture than profit making, they are much concerned about quality of products and services.

In the United States, for example, the majority of historic sites are owned and managed by nonprofit organizations. This list includes the Alamo, Texas; Williamsburg, Virginia; Mount Vernon, Virginia; Mystic Seaport, Connecticut; the Polynesian Culture Center, and Aloha Week festivals, Hawaii; Mardi Gras, New Orleans; the Biltmore Estate, North Carolina; Ballingrath Gardens, Alabama; and all Nature Conservancy sites.

Generally, the nonprofit sector is one in which charges for products and services are returned for operating expenses but not for the accumulation of wealth. Typically, revenues are obtained from gate receipts, admissions, donations, and foundation grants. Some nonprofits receive monies primarily from members.

Another kind of nonprofit group directs its interests toward improvement of a tourism sector. In the United States, more than fifty nonprofit organizations are involved in tourism (Weaver 1991). These include: the Air Transport Association, the American Automobile Association, the American Bar Association, the American Hotel and Motel Association, American Sightseeing International, the American Society of Travel Agents, the Conference of National Park Concessionaires, the Council on Hotel, Restaurant and Institutional Education, the Highway Users Federation, the National Campground Owners Association, the National Tour Association, the Travel Industry

Association of America, and the Ecotourism Society. These organizations provide members with technical assistance, news on topics related to their interest, and opportunities for information exchange. Also, each province of Canada has a travel industry association. Most festivals are sponsored by nonprofit organizations.

In Scotland, the National Trust, a voluntary organization, owns and operates many castles, mansions, gardens, historic sites, battlefields, birthplaces of famous Scots, and examples of vernacular architecture—all of significant value to visitors (Scottish Tourist Board 1994). Many other nonprofit-run developments in that nation serve travelers, such as museums, golf courses, artistic events, galas, tennis courts, and highland games. Historic Scotland operates 330 properties, including the popularly visited Edinburgh Castle and other attractions throughout the nation, from Dumfries and Gallaway in the south to Shetland in the north (Scottish Tourist Board 1994). These attractions assist greatly in extending the tourism season, spreading the distribution of visitors, and providing cultural enrichment to visitors.

Governments

A very important third sector of developers and managers of tourism establishments and programs is the public sector. In many countries, governments at the national level own and operate airlines and travel agencies. Tourism in Japan, for example, has three major divisions within its Department of Tourism—Planning, Travel Agencies, and Development (Japan National Tourist Organization 1991). The government funds resort planning, subsidizes construction, and aids promotion. Prefectural governments are responsible for roads, parking, campgrounds, and public utilities.

Worldwide, the public sector is a huge promoter of tourism. Billions of dollars are spent by states, provinces, cities, and national governments on promotion.

At the city level, the public sector is a major provider of the basic infrastructure used by tourism. Street systems, lighting, water supply, waste and sewage disposal, and police and fire protection are essential elements developed and managed by the public sector. In addition, this sector controls many amenities created mainly for local residents but significant to travelers as well, such as museums, outdoor theaters, performing arts venues, park and recreation areas, and festivals.

In summary, all three sectors (governments, nonprofit organizations, commercial enterprise) are involved in all five components of the supply side of tourism development. But, because the individual players have their own resident-oriented mandates, philosophies, sites, and practices, the mechanisms for cooperation and collaboration among them have been very slow to develop.

As McNulty (1994) has pointed out, these three sectors have changed in recent years. They are less sharply defined, they are seldom under local control, and their leadership has become institutionalized. Historic sites, for example, today may be owned and managed by public park and historic agencies, nonprofit historic societies, or even private businesses. Today, a great amount of tourism investment comes from outside rather than local sources. The individuals who provided local leadership in the past are now being brought into large organizations where continuity can be maintained. All these changes are modifying the patterns of who plans, builds, and manages tourism. All present a challenge to local communities who seek to guide their own destinies. Local governments, organizations, and the commercial sectors now must review these changes and discover new roles for themselves in regard to tourism. No longer can they isolate themselves.

Outside investors in tourism must be aware of local impacts and make sure their development is properly integrated into the community. Conversely, local residents must identify their community values and articulate them to foreign investors. Local governments, once solely focused on resident needs, now must plan and administer for visitor populations as well. The commercial sector of tourism is no longer isolated, but is an integral part of the decision making required of governments, nonprofit organizations, and especially local citizens.

Increasingly, governments are dropping many of their programs for tourism (Walker and Smith 1995). Many public services, such as railroads, canals, bridges, tollways, and communication systems are being privatized. Governments pressured by other demands from tax revenues are turning to the private sector (both nonprofit and commercial) for help. Usually private-sector development is preferable; it can reduce tax burdens on the populace and respond more quickly and efficiently to fluctuations in market demand. Examples of this change in sector development responsibility include: the Hong Kong Cross Harbor Tunnel; the Shajiao 'B' Power Station, China; the Toronto airport; North South Highway, Malaysia; and the Dartford Bridge, United Kingdom. Developers of tourism everywhere need to be on the alert for new business opportunities resulting from changes in government policy.

CONCLUSIONS

Although others may influence development, it is within the power of communities and destinations to make decisions on development of all five components of the supply side—attractions, transportation, services, information, and promotion. This development must satisfy the wants and needs of visitors. Planners and developers of the supply side need an understanding of market characteristics so that they can best determine what should be developed. A major goal of development is to attain a balance between supply and demand.

Contrary to popular opinion, tourism is not solely under the control of the commercial enterprise sector. Governments and nonprofit organizations invest, build, and manage tourism development as well. The challenge to create a more smoothly running tourism system requires new communication and cooperation among all these sectors.

An important component of this discussion is that both demand and supply are very dynamic. Changes in markets affect all parts of the supply side. Deterioration or improvements in supply—attractions, transportation, services, information, and promotion—change market appeal and visitor rewards.

Finally, all tourism development must be integrated into community and destination development. Tourism is not an isolated overlay; it is an integral part of every aspect of local development.

For tourism to succeed, all components of the supply side must function together, such as for the annual Gasparilla Invasion Festival, Tampa, Florida. Visitor housing, food services, transportation, descriptive guidance, and advertising are interrelated parts of making the festival a success. (Photo courtesy Tampa News Bureau)

C|H|A|P|T|E|R| |5|

Attractions: First Power

BACKGROUND

Washington, DC, as the seat of American government and shrines, draws many thousands of visitors who carry out personal, professional, political, and business activities. Houston, as a trade center, lures thousands of visitors who conduct business and participate in conferences and conventions. The unique land characteristics of Hawaii Volcanoes National Park have made it one of the most popular parks in the world despite the 2,600 miles of ocean that separate it from the U.S. mainland markets. The appealing forests and waters of various youth camps encourage young people to take part in outdoor experiences.

Attractions are the lifeblood of tourism everywhere. Kuala Lumpur, capital of Malaysia, and its associated attractions—the Shah Alam mosque, the Thean Hou Temple, and Lake Titiwangsa—bring increasing volumes of visitors to Southeast Asia. Trafalgar Square, Big Ben, and Buckingham Palace attract throngs of visitors from all over the world, as do the exotic features of other world cities. The countrysides of Italy, Ireland, and Spain appeal to increasing numbers of travelers. Other travelers are lured by the historic haunts of famed writers and poets such as Ezra Pound, Ernest Hemingway, and Lord Byron. Today's cruises are noted as much for their portside attractions— for example, the Balearic Isles, Genoa, Hamburg, Bergen, Stockholm, Copenhagen, Hokkaido, and Tasmania—as for the extravagant shows and food on board.

Despite their diversity, these places have in common a "nonhome" appeal (Lew 1994). All are targeted for purposeful, rather than haphazard, trips. Because of the pull they exert on the traveler, all can be classified by the generic term *attraction,* the main power that drives tourism everywhere.

An Ancient Concept

Throughout time, mankind has had a burning desire to experience the exotic, and just as quickly as means could be made available this desire was fulfilled. The ancient Greeks traveled within their own land and to nearby countries for trade, cultural enrichment, and better health. (The local spas were well equipped with physical fitness programs.) The Romans were lured by the sea, the countryside, and the cosmopolitan pleasures offered by the cities, including theater and gladiator combat. Then as now, shrines, religious festivals, and athletic competitions were prominent attractions.

Through the ages educational value has been attached to travel. In Shakespeare's *Two Gentlemen of Verona* it is well promoted:

Panthino: Some to the wars, to try their fortune there,
    Some to discover Islands far away,
    Some to studious Universities.
    For any or for all of these exercises
    He said that Proteus, your son, was meet,
    And did request me to importune you
    To let him spend his time no more at home,
    Which would be great impeachment to his age
    In having known no travel in his youth.

Antonio: Nor need'st thou much importune me to that
    Whereupon this month I have been hammering,
    I have consider'd well his loss of time
    And how he cannot be a perfect man
    Not being tryed and tutor'd in the world.
    Experience is by industry achieved
    And perfected by the swift course of time.
    Tell me whither were I best to send him.

(Act 1, Scene 3; Neilson, Allen, and Hill 1623/1942, 31)

In the eighteenth and nineteenth centuries no English student of quality completed his education without taking the "grand tour." Undoubtedly the first package tour in history, this extensive trip included the highlights of France, Switzerland, Italy, Germany, and the Low Countries. The beauty of foreign lands profoundly influenced the literature of Rousseau, Shelley, Byron, and Ruskin, and the paintings of Turner. In this period, landscape art was doing an equally powerful job of opening the eyes of masses to the pleasures of foreign travel.

Travel has been significant in societies for centuries. The cultural and trade attractions of Greece and Rome were very important in ancient times and continue to provide enjoyment and cultural understanding today. (Photo of Irodus Atticus Theatre, courtesy Union News Photo, Athens)

Attraction Constants

An examination of two centuries of tourism in the United States shows that many present activities are similar at least in name to those of yesterday. Early accounts reveal the popularity of such then-illegal pastimes as "dice, cards, quoits, bowls, ninepins . . . Shuffle Board" (Dulles 1964, 6). Good food, wine, and whiskey have been accompaniments to many recreations throughout the history of America despite Puritan claims to the contrary. Strolling, people-watching, attending concerts, and using "gouff clubs" were urban pastimes of the early 1700s, as were hunting, fishing, country fairs, target practice, pleasure boating, ice carnivals, skating, hockey, horse racing, and other attractions requiring travel.

Scenic excursions from town to town, the antecedents of our present pleasure drives and cruises, were common two centuries ago. Cockfighting was a popular sport in New England and the South. Even the modern zoo had

its counterpart as early as 1733, with camels, lions, polar bears, monkeys, and elephants on display in the United States (Dulles, 1964).

The names may have persisted for some activities, but their functions are now expressed in different ways. Health spas have all but abandoned the mineral water cures that were famous at the turn of the century, incorporating instead a variety of health, wellness, fitness, and weight loss facilities and programs. Fishing, once simply a matter of angling, now includes the use of plane-to-boat communications and sonar to spot fishing beds. Vacation homes, previously restricted to personally owned cottages, now include condominiums in high-rise buildings or row houses as well as time-sharing options. Ship travel, formerly an upper-class means of transoceanic transportation, is now dominated by an elaborate floating resort concept, with on-deck recreations and epicurean delights.

For the developer, perhaps the most important changes have been the proliferation of activities and the broadening of clientele. A simple activity such as camping has myriad offshoots, including wilderness camping, backpack camping, boat camping, tent camping, and RV camping. Golfing is no longer the exclusive sport of the wealthy. Sightseers include a wide range of individuals, from the student hitchhiker to the multimillionaire with his private island on the opposite side of the earth. "What had once been largely restricted to the genteel members of society had become the property of the people as a whole" (Dulles 1964, 396).

This history of U.S. travel is not unlike those elsewhere in the world. It shows changes in use of attractions and helps demonstrate that most change involves the manner in which activities take place. The only limits are those of imagination and creativity, which are largely the responsibility of planners, designers, and developers at the community and destination level.

Attractions Relocated

Because of changes in transportation technology, more of today's attractions are to be found in mass complexes at travel nodes rather than scattered along roadsides. Variations in the placement of attractions are of special significance to designers, planners, and developers, especially at the community level.

A major change in the location of attractions came with steamboat and railway travel in the late 1800s. For the first time, attractions were clustered at travel termini and exchange points, usually at communities. In the Great Lakes region, for example, resorts with Victorian grandeur, promenades and all, flourished at port cities and points where rail lines touched beautiful lakes and lumber towns. The resort attractions were tightly grouped about the hotels, and guests strayed only as far as horse, carriage, or canoe would allow. This same pattern prevailed throughout the world.

The automobile age initiated another major change, enabling travelers to reach thousands more attractions along roadsides rather than merely at steamboat and railway termini. Salesmen had access to rural areas and small towns, and pleasure travelers could reach more parks, mountains, and beaches. The proliferation of highways and the comparatively slow speed of automobiles fostered hordes of roadside attractions, such as snake farms, shell collections, mystery houses, and shops filled with trinkets and curios.

Whereas the beginning of the automobile era fostered

Medieval Toledo, the capital of Spain in 1085 and now a Spanish national monument, is a major European attraction. Note relics of past cultures in Moorish and Gothic architecture.

Petoskey, Michigan, was a major resort destination for travel markets in the United States in its early history. This photo, taken in the summer of 1882 at the railway depot, demonstrates the popularity of this destination during the steamboat and railroad era. (Photo courtesy Chicago, Grand Rapids & Indiana Railroad)

The category of touring circuits continues to grow on land and sea. Motorcoach tours, travel in private automobiles, and cruises offer a variety of interesting experiences. Illustrated here is the Patagonian Cruise, Natales, Chile. (Photo courtesy Amrit Kendrick)

Equally important is the category of longer-stay tourism. Visits to resorts provide an abundance of activities and are often segmented by traveler interests. Typical of warm and sandy beaches of the world are resorts such as at Waikiki, Hawaii.

an explosion in mobility that allowed penetration into untouched areas, the modern expressway and jet travel have returned tourism development to a tighter clustering of attractions. *Place,* the essential characteristic of an attraction, has now become far more important than in the days of shoepack and donkey. "The history of tourism clearly indicates that the environment of places has contributed to the birth and progress of tourism" (Mathieson and Wall 1982, 94).

ATTRACTIONS CLASSIFIED

The abundance and diversity of travel attractions seem to defy description and certainly complicate their design and development. People travel worldwide to shop in department stores, to participate in international conferences, to do business with multinational firms, to visit shrines, to view contemporary and historic gardens, and to participate in cultural events.

One useful means of classification is to divide all attractions into two general market-driven classes: touring circuit and longer-stay. By *touring circuit* is meant those attractions visited on a tour and for a comparatively short time. Because tourists are flowing through the area, the development of such attractions requires specific resources, designs, and operations for the successive groups of tourists that will visit throughout each day.

These travelers will visit many locations in the period between leaving home and returning. On the other hand, attractions for *longer-stay* use require resources, design, and operations for groups of people who will stay for more than a brief visit. For example, a historic site can be toured in a relatively short time, perhaps an hour or so. However, travelers who use vacation homes or resorts or attend conferences stay in one part of a destination zone for a longer time. A resort visitor will fish, boat, or swim in one general area day after day, whereas a touring-circuit traveler will attend an outdoor drama but once.

Table 5-1 lists attractions according to these two categories. This list can help the developer recognize the difference between attraction objectives for these two major classes. Figure 5-1 illustrates the degree of dependency of attractions upon natural and cultural resource foundations.

Although these two types of attractions may seem to have little in common, a development theme may be threaded throughout all. Frequently, attractions are united by a common geographic area, even though they function differently. New England's "roadside scenic areas," "outstanding natural areas," and "historic buildings and sites" are frequently linked together on visitor tours. Winter and summer "resorts," "camping areas," and "fishing and hunting areas" are often clustered together. This pattern suggests the need for collaboration, or at least cooperation, among the planners and developers of

TABLE 5-1. Attractions Classified by Length of Stay

Touring circuit attractions	Longer-stay attractions
Roadside scenic areas	Resorts
Outstanding natural areas	Camping areas
Camping areas	Hunting and water sports areas
Water touring areas	Organization camp areas
Homes of friends or relatives	Vacation home complexes
Unusual institutions	Festival and event places
Shrines and cultural places	Convention and meeting places
Food and entertainment places	Gaming centers
Historic buildings and sites	Sports arenas and complexes
Ethnic areas	Trade centers
Shopping areas	Science and technology centers
Crafts and lore places	Theme parks

Source: Gunn 1988, 42.

separate attractions, who must nonetheless maintain the individuality of each.

Certainly, the amount of time travelers budget for visits to attractions has much to do with how they are created. Planners of motor coach tours usually cannot include theme parks in their itineraries because they take too much time. Short-term attractions require quick and easy circulation, but focused attractions can be designed for more casual, exploratory use.

Because economic motives are so strong, the greater the aggregation of both touring circuit and longer-stay attractions, the better (Figure 5-2). Groupings provide business support for the services desired by travelers, such as hotels, food, and entertainment, but on the other hand they also trigger concern over saturation and the environmental consequences of too many visitors. The principle of grouping has helped foster a planning and development focus on host communities and destinations.

ATTRACTION COMMONALITY

For the planner and developer, the complicated mass of attractions can be baffling. On the surface, no consistency is apparent. How can a visit to the Great Smoky Mountains National Park and a safari in the Kalahari Desert have common features? For planning and development purposes on the supply side, it is necessary to know more about the characteristics such itineraries share. Recognition of six common factors is important for revising existing or establishing new attractions.

All attractions must be well understood by visitors. This function can be assisted by well-designed interpretive centers and qualified tour guides, such as provided at the Tower of London.

KINDS OF ATTRACTIONS	DEPENDENCY UPON NATURAL RESOURCES	DEPENDENCY UPON OTHER THAN NATURAL AND CULTURAL RESOURCES	DEPENDENCY UPON CULTURAL RESOURCES
TOURING CIRCUIT:			
Roadside scenic areas	◉	○	○
Outstanding natural areas	◉	○	○
Camping areas	○	◉	○
Water touring areas	◉	☆	☆
Homes: friends/relatives	☆	◉	☆
Unusual institutions	○	◉	○
Shrines, cultural places	☆	★	◉
Food, entertainment	☆	◉	○
Historic bldgs., sites	☆	○	◉
Ethnic areas	☆	★	★
Shopping areas	☆	◉	☆
Crafts, lore places	☆	★	◉
LONGER-STAY:			
Resorts	◉	★	☆
Camping areas	◉	★	○
Hunting, water sports	◉	☆	○
Organization camps	◉	☆	○
Vacation home complexes	◉	○	○
Festival, event places	○	◉	◉
Convention, meeting places	☆	◉	☆
Gaming centers	☆	◉	○
Sports arenas, complexes	○	◉	○
Trade centers	○	◉	★
Service, tech. centers	○	◉	○
Theme parks	☆	◉	○

◉ Highly Dependent   ★ Dependent   ☆ Somewhat Dependent   ○ Low or No Dependency

Figure 5-1. *Resource dependency of attractions. Most attractions are founded, in varying degrees, upon natural and cultural resources. A few have weak relationship to these resource foundations.*

Easy Comprehensibility

Planners and developers of attractions must ensure that the attractions are readily understood by those who use them. Every attraction should provide the user with the information, or perhaps skills, necessary for fullest participation. For example, in sports, many resort areas offer instruction in marksmanship, swimming, golf, water-skiing, snow skiing, fishing, or hunting. Many national parks are staffed by professional nature interpreters who tell of the area's unique flora and fauna. Historic, archaeologic, and other heritage sites and restorations not only are labeled with descriptive information but are interpreted by guides, taped narrations, and colorful displays. If the visitor cannot understand the attraction, he loses interest, and it ceases to be an attraction.

The developer must exercise restraint so that the mechanisms that foster communication between visitor and site do not overwhelm the visitor. Sometimes visitor centers, exhibits, and signs overpower the attraction by their sheer mass and number. There is a delicate balance between allowing the visitor to gain adventure from the attraction through his own initiative and choking the visitor with information.

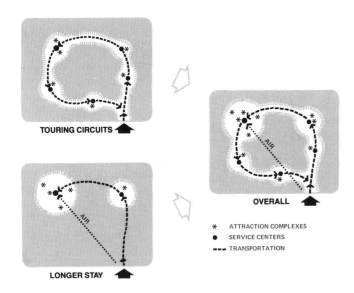

TOURING CIRCUITS

LONGER STAY

OVERALL

* ATTRACTION COMPLEXES
● SERVICE CENTERS
--- TRANSPORTATION

Figure 5-2. *Combined development of tourism. Communities have the best tourism economy when serving both touring circuit and longer-stay travelers. This combination may also cause the greatest stress unless well planned.*

Basis in Environment

The environmental foundation for an attraction has many important implications. Even if the Sphinx were removed to the Smithsonian Institution, its roots would still be in the Sahara. Every attraction has powerful place characteristics, both by physical location and by association. For all attractions that association is important; for some it assumes an especially strong significance. It is doubtful, for example, that husbands and wives ever forget the place of their honeymoon. "Love of one another is linked to love of place" (Shepard 1967, 33).

The fearless wildlife of the Galapagos lies off the coast of Ecuador, and no amount of mental exertion could place it conceptually in Times Square. Any attraction is swept up within its native climate, other natural influences, and, especially, the manmade influences surrounding it. The transplanted London Bridge remains integrally related to London even though it has been relocated to Arizona. Even "siteless" attractions, such as gambling, parades, and pageants, occur in some kind of setting. In the design and development of attractions,

Environmental setting and anchorage to place are important characteristics of attractions. Much of the appeal of the South Pacific Islands, such as Moorea, is because of their dramatic mountain scenery, Polynesian cultural foundations, French ethnic influence, and crystal clear waters.

therefore, the implications of environment cannot be ignored.

In an age of anonymity, it is important to seek out and preserve qualities that distinguish one place from another. Too often the inherent amenities and natural features that all places possess are sacrificed for the sake of economy or convenience, resulting in a loss of identity (Johnson Johnson & Roy, Inc. 1973).

Both environmental assets and constraints are important to the design and development of tourism. Abundant and high-quality surface water, vegetative cover, wildlife, and land relief (hills, valleys, mountains), as well as a climate favorable to outdoor recreation, are important resources upon which tourism may be developed. Polluted water, cold water (short season use), a desert climate, sparse or diseased wildlife, and a climate neither cold enough for winter sports nor warm enough for relief from northern climates have limited tourism potential. The suitability of a place and its natural assets for tourism development is further enhanced when transportation access is available and a service city is nearby.

In the last decade, market segments interested in history and cultural backgrounds have increased greatly. This demand factor has brought historic sites and buildings, ethnic arts and crafts, festivals, shrines, and manmade spectacles (e.g., space vehicle launching) into the category of resources for tourism development.

These many natural and cultural resources form the foundation upon which tourism development can be considered. It must be emphasized, however, that these resources are specific to locations and are not equally distributed over a single region or country.

In the course of time, a landscape, whether of a large region such as a country or a small locality such as a market town, acquires a specific culture- and history-conditioned character, which commonly reflects not only the work and the aspirations of the society presently in occupancy but also those of its precursors in the area. Whitehand (1992) makes the case that the most important issue of urban landscape management is how well new developments, such as tourism, fit into the environment and become an integral part of existing place qualities. Especially critical are areas that reflect historical acculturation.

The physical form, the natural and cultural assets, and the quality of the environment are as essential to local living as to visitor experiencing and enjoying. Sensitivity to local needs is requisite to all planning and development of community-based tourism and its attractions.

From the community perspective, certain common needs are especially applicable to tourism development. These have been identified by McNulty and Page (1994) as *opportunity, equity,* and an *environment that supports life.* It must be remembered that tourism cannot be considered as merely an economic overlay; instead it must be a part of the opportunities for economic and social life of the residents. Similarly, the amenities for residents must be shared by visitors. Equity refers to the feeling of residents that they own and control their community. This factor will not tolerate discrimination in any form.

Owner Control

An important ingredient for the success of any attraction is effective ownership and management by an individual or group, either public or private. Their specific policies and practices may vary. Commercial owners create attractions for business reasons; government and nonprofit owners create attractions that fulfill social objectives.

Many countries have traditionally looked to government for the development of attractions that are based primarily on natural resources. This is based on the belief that only governments can supply wholesome and worthwhile attractions. In recent years, however, the Disneylands and Williamsburgs have demonstrated that social values can be upheld by attractions designed and managed by the private sector.

Planners and designers play a critical catalytic role in balancing ownership policies with the needs of visitors. For example, the promoter of a resort hotel complex may wish to take full advantage of its linkage with a beautiful beach and water resource. He may consider a site directly on the beach to be optimal. However, the planner may emphasize the finite amount of beachfront and favor clusters of structures farther back from the beach that allow open spaces between them to maximize views and access to the beach. It is planners' role to help developers reach their objectives and at the same time protect both visitors' interests and environmental assets. Throughout the process, designers can guide more satisfying and attractive developments, no matter what ownership policies prevail.

Communities that are looking toward expanded tourism based on government-owned and managed areas such as parks must identify all policies concerning visitor use. The times the attraction is open, its fees and charges, seasonality, and all regulations for public use have great bearing on the development relationship to the community. Obtaining guidance on these matters requires close cooperation between the relevant government agency and community tourism leaders.

Magnetism

By definition an attraction is magnetic; it draws people. This concept, that an attraction is defined by its pulling power, is antithetical to the beliefs of many for whom an attraction comes into being merely by the owner's declaration and construction. But the true test is pulling power.

The concept of magnetism has two corollaries of con-

The manner in which visitors may use an attraction is subject to the policies of ownership and management. Developers and promoters of tourism must consider these policies for all tourism planning and development. Hacha Falls, Venezuela, as a tourist attraction, is subject to control by managers of Canaima National Park.

cern to the developer. First, magnetism exists in the eyes of the visitor, and each visitor has unique interests and preferences. Second, magnetism is also a product of the design, development, and managerial operation of an attraction. Designers can create magnetic attractions based on given environmental assets and visitor interests. Emphasis must be placed on the requirement for an attraction to meet the needs of a specific market segment or several segments at one time. When a designer brings market interest and resource potential together to create an attraction, its success is assured.

Capacity to Satisfy

Another corollary of pulling power is visitor satisfaction. A successful attraction is rewarding to the participants. Of course, attendance figures alone do not reveal the depth of user satisfaction. This is the major challenge in designing and establishing attractions. If the visitor leaves feeling disappointed, uninterested, or even defrauded, the attraction may have succeeded in attracting but not in carrying out its complete function. If a developer is to produce successful attractions, his plans and establishments must elicit user satisfaction. To achieve this objective a thorough understanding of market segments is required.

Result of Creation

Every attraction today is created. This statement may seem fatuous, especially in the face of such natural wonders as the Grand Canyon in the United States or Giant's Causeway in Ireland. Nevertheless, in the context of modern tourism, even these most compelling places do not become true attractions until they are provided with access, lookout points, parking areas, interpretation programs, and linkages with service centers.

One can think of this concept in the negative as well. Our ability to alter the environment is so facile today that choosing not to change a natural feature is itself an act of creation. However, seldom can an attraction be merely a locked-up resource. Fires, earthquakes, floods, insects, and disease change or destroy natural features, and management control must be exercised even only to preserve them.

Along with the power to create attractions must come the responsibility to protect resources. These are two sides of the same coin. One can protect fragile sites by limiting structural development and restricting visitor use. Meanwhile, one can develop "hardened" sites nearby for facilities and services.

Consideration of these characteristics of attractions

By definition, attractions must be magnetic in their action on visitors, not just in the opinion of an optimistic developer or promoter. Access to Pagsanjan Falls, a popular attraction in the Philippines, is by a dugout canoe trip from a private resort hotel.

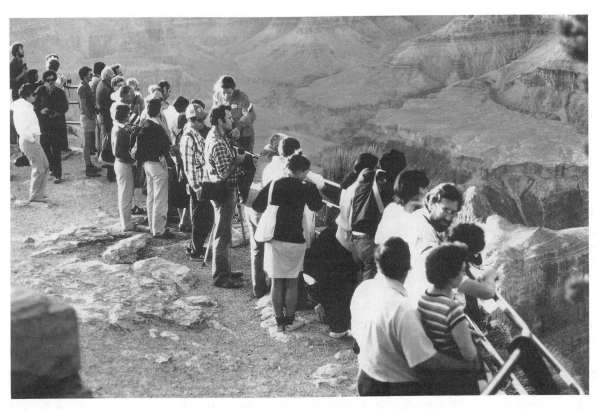

The ultimate measure of an attraction's success is how well it provides satisfaction to visitors, such as these travelers waiting for the spectacular sunset views of the rugged rock formations at Grand Canyon National Park.

Not only commercial theme parks are created; so are national parks and preserves. Creativity of policy and planning have set aside natural treasures such as the "peace park" of adjoining Waterton Lakes National Park, Canada, and Glacier National Park, United States. Land use design and management for resource protection and visitor use are acts of creativity.

can help communities and destinations establish places that are key to development of other components for best functioning of tourism.

AUTHENTICITY

For some, the idea of "designed," "created," and "developed" attractions may smack of arrogance or even fraud. Travelers around the globe complain about gross misrepresentations. There have been instances where jewelry, crafts, dances, and costumes were identified as having genuine ethnic origins but turned out to be fakes. Festivals are sometimes falsely advertised as based in historical fact. "Tourism has been accused of being 'culturally arrogant' for manipulating the traditions and customs of people to make tourist experiences more interesting and satisfying" (Mathieson and Wall 1982, 173).

But how can one define a "true" experience? What is now one of Hawaii's greatest appeals, its fluid patois in song, came not so much from the native Polynesian music as from New York's Tin Pan Alley songwriting of the early 1900s. Using this nonnative genre, composers wrought an amalgam of the special sights, sounds, smells, and language of Hawaii that elicits visitors' heartfelt emotions, as would never be possible from the non-

melodic grunts of pure Polynesian origin. Who would say that these very particular and pleasurable feelings millions of visitors associate with Hawaii are false?

When a developer reconstructed old Fort Michilimackinac in northern Michigan in the 1960s, he faced a major decision regarding authenticity. The original structures, erected in 1763 and long rotted away, had been built on the ground or on local stone rubble. The developer-historian, Dr. Eugene Peterson, made the policy decision that all the 1763 structures would be built on concrete for durability, but for period aesthetics this foundation would be faced with the original stone. This method would present the buildings in their true setting, and the $3 million invested would not have to be replaced in a few years. Millions of future visitors will appreciate this policy decision.

Historic reuse is an attempt to have the best of both worlds—the essence of an earlier era without the high cost of maintaining a museum. Pierre Berton (1985, vii) has commented about Canada's Main Street program: "It's no use preserving a building unless a use can be found for it. It's no use preserving a streetscape if it is not economically viable." For such buildings to function as travel attractions, new air conditioning, heating, plumbing, rest rooms, and electrical systems must be incorporated. All these modifications test the skill of designers, who must ensure that the spirit and drama of the "real" thing is not spoiled.

The issue of authenticity of tourist attractions also relates to the ethics of promotion. If buildings and sites are purported to be something they are not, the public may be disappointed or even angry. Travelers today are more sophisticated than many developers and promoters realize. It is important to be careful when using descriptors such as "the original," "the real thing," "the exact place where this event took place," or "handmade." Honesty in advertising demands that such phrases as "a replication," "near the place of," or "manufactured" be used where appropriate. The difference between a documentation and a drama needs to be explicit. Drama is an art form and can be an important attraction adjunct when presented honestly.

Restored historic buildings often suffer from inappropriate landscapes. Whereas mass use may necessitate such features as hard surfaces for walks, the designer can often utilize period designs and materials that minimize wear and tear. Research into the landscape materials of a period is as important as the study of its architectural detailing.

TRIPARTITE ATTRACTION CONCEPT

The study of tourist attractions in the context of visitors and settings reveals elements and relationships that may be useful in the design of future attractions. The concept of an attraction being composed of three important functional parts provides a useful framework for examining these elements and their relationships (Figure 5-3) (Gunn 1965).

Nucleus

The prime element of an attraction, its raison d'être, is the *nucleus*. For a waterfall, the nucleus is the sound and view of falling water; for a mountain, it is the peak; for an area rich in historical significance, it is the landscape or building. In the design of attractions the nucleus must authentically represent its foundation and be of a type and quality to match or surpass the images held by tourists.

If the nucleus is a fragile or rare resource, extreme care must be taken in planning for visitors, especially for large numbers. The public may be held at the edge of a prime resource feature and allowed to experience it vicariously through interpretive lectures, presentations, pageantry, simulation, exhibits, overlooks, or specially designed transportation for sightseeing. Methods must be used that prevent physical contact that would be damaging to the environment.

Visitor enjoyment and appreciation of many travel destinations are now severely diminished—for example, by helicopter tours over the Grand Canyon, USA. Air, noise, and aesthetic pollution can damage an attraction nucleus if not controlled.

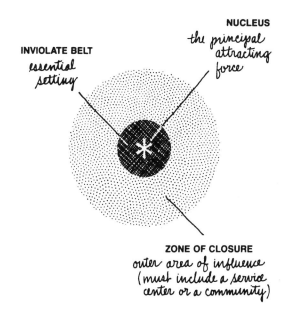

Figure 5-3. *Tripartite attraction concept. In attraction development, the planning and development of three spatial zones should be considered—the nucleus, the inviolate belt, and the zone of closure (which includes one or more communities).*

Inviolate Belt

The function of attractions depends equally on the setting, or *inviolate belt*. The visitor can reach a feature (nucleus) only by passing through some buffer space, which may be small or large, brief or of extended duration. Physio-psychological conditioning is the function of this space. It is the frame for the feature.

A person's mindset or anticipation of an attraction has much to do with his reception and approval when the nucleus is reached. Without a doubt, the inviolate belt has a much more powerful function than has been previously attributed to it. No nucleus can be without it. Its creation requires special sensitivity and creativity on the part of the designer. Even the areas around "siteless" attractions such as theme parks and gaming casinos must be planned with supportive and compatible tourism supply functions.

The inviolate belt can be very difficult to incorporate into the overall design of an attraction. Others may own properties surrounding the nucleus, and they often have quite different land use purposes and policies. For example, a historic building is often surrounded by contemporary structures with no visitor functions. Every effort should be made to consider the design and development requirements of the inviolate belt surrounding an attraction nucleus.

Zone of Closure

Especially important to all tourism developers is the third aspect of an attraction, its surrounding area, or *zone of closure*. Within this zone must be found one or more service centers as well as transportation linkage between the service centers and the attraction. Service centers contain the business places and community services needed by traveler—lodging, food, entertainment, car service, communications, banking, information, and retail purchases. No matter how remote the attraction, some service center must be available and accessible for the attraction to function.

This principle is demonstrated by the relationship between the service centers of Gatlinburg and Cherokee and the attractions of the Great Smoky Mountains National Park (Figure 5-4). The National Park Service protects the several attraction nuclei and provides visitor interpretation. The nuclei are surrounded by an inviolate belt of a complementary landscape setting. The many services desired by travelers are provided in the nearby cities and linked to the park by major highway corridors.

The zone of closure concept suggests a need for greater development and jurisdictional cooperation between the developer and community than generally exists today. The interdependent functions of the many sites within the zone are not always clear to their numerous owners. The professional planner or landscape architect can assist all owners and developers in understanding the personal benefits of applying this concept. Developers of attractions can exercise resource protection policies more effectively without having to provide services. Businesses that do provide services can become more successful by understanding the market trends of visitors to the attractions. Finally, traveler satisfaction can be enhanced by an integrated plan for service centers, attractions, and transportation linkages.

COMPLEXES

Attractions grouped into larger complexes thrive better than smaller, isolated ones. This principle has been fostered partly by the dominance of expressway and jet travel, which supports the clustering of attractions at termini. But more likely it is the result of travelers' demands for more recreational and business opportunities at destinations. Detailed examination of natural and cultural resources can reveal their potential for incorporation into larger complexes. For example, important forest, wildlife, or historic sites may be prime areas for attraction complexes with scenic, hunting, or historical themes. A study of markets can reveal the extent to which a designer can consolidate compatible complexes, even those that have different owners.

An excellent example of a historical complex is the Museum of Appalachia, located in Norris, Tennessee (Ralston 1996). Instead of only one historic building in an isolated setting, this rich demonstration of the Appalachian subculture of generations ago comprises more than thirty buildings on 65 acres of land.

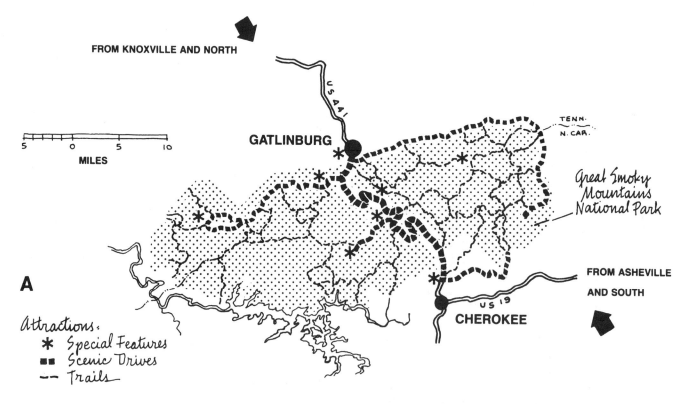

Figure 5-4. *Community–park relationship. Interdependency between service communities and natural resource attractions is shown by this plan of Great Smoky Mountains National Park and Gatlinburg and Cherokee, Tennessee.*

The isolation of the mountain people of Tennessee, West Virginia, Kentucky, Virginia, and North Carolina produced a self-sufficient regional culture. The museum demonstrates how these people developed their own shelter, food, furniture, tools, and entertainment directly from the land resources. The historical context ranges from the 1700s through the 1800s. Museum entrepreneur John Rice Irwin, many years ago, was concerned about the gradual decline of this culture. In his lifetime he had witnessed its modernization, and he vowed to preserve and interpret it for future generations. Today several hundred thousand visitors have the opportunity to gain insight into this very special American cultural heritage.

Many national parks are large attraction complexes; they provide opportunities for viewing scenery, photographing natural resources, hiking, and horseback riding, as well as for social activities related to the park, such as camping and visiting interpretation centers. The Red Rock National Park of China exemplifies an attraction complex (Yu 1995). It encompasses a variety of individual attractions, including the red rock features, evergreen forests, water landscape features, and historic structures.

A major complex is more efficiently engineered and controlled than many smaller attractions widely dispersed. The costs of water supply, sewage disposal, electrical power, and roads are reduced in a single complex. Police protection, fire protection, and management control is likewise more efficient in a more compact setting. As public park and theme park managers have experienced, it is much easier to manage crowds in a single area with one entrance than in many areas far apart. And, best of all, a complex offers the visitor a much more complete and enjoyable experience.

When a study of markets and resources suggests that various visitors' interests may conflict, attraction complexes should be segregated. Incompatible elements should be designed and managed independently. For example, park managers have successfully separated areas for day users from those for longer-stay users. Tent campers have been accorded different sites from RV campers. Historic site visitors have been separated from beach and boat users. And wilderness buffs have pursued their interests in places unoccupied by gregarious park users whenever policies and plans have recognized the need for segregation.

Recently, the issue of more research into attractions has been raised. If more were known about this phenomenon, perhaps designers, planners, and developers would have better guidance for the future. For example, Lew (1994) examined many studies of attractions in an attempt to identify consistencies. Lew recommended that attraction typologies be assessed on the basis of four measures: ideographic, organizational, cognitive, and cross-perspective. This and other studies have endorsed the concept of attractions as a complicated but most important component of all tourism development.

CONCLUSIONS

The main conclusion to be drawn from this discussion is the recognition of the great power of tourism's major supply-side component—attractions. Attractions drive all other components. Improved and new attractions represent the greatest opportunity for community and destination development. When attractions are in place and available to visitors, the planning and managing needs for the other components of supply—transportation, services, information, and promotion—become apparent.

These simple statements, however, belie the many complications of attractions and the need for great care in their selection, development, and management. An understanding of the role, characteristics, classifications, and concepts for development of attractions is essential for tourism planners.

Because attractions are inextricably of the land, their planning, development, and management are greatly dependent on all other trends in local land use. Perhaps the greatest threats, especially in undeveloped regions of the world, come from other demands on resources. As the world population grows, societal and economic demands on resources escalate. The next thrust of mineral, oil, and forest exploitation will be in those very regions with greatest tourism potential. The temptation for nations to export and sell these resources is very great. However, once the forests and wildlife are depleted, mining has stripped the landscape and polluted the waters, and local cultures have been destroyed, these resources for tourism are gone and cannot be replaced.

Local communities face many difficult decisions as they try to balance the future use of their economic and cultural foundations, especially for tourism.

CHAPTER 6

Destination Development

DESTINATION ZONE DEFINED

The analysis of the parts of tourism, especially the demand and the supply sides, provides a foundation for creation of community tourism development. The interdependence of the five components—attractions, transportation, services, information, and promotion—highlights the need for cooperation and collaboration. The extremely significant role of attractions heightens the demand for careful resource planning and management. And the huge array of decision makers on development compounds the task of expanding tourism at the community level. These are the factors that every community must deal with if it anticipates tourism to meet today's standards of success and quality.

Places that represent *here* for the local resident represent *there* for visitors from another origin. The characteristics of "here," if they are to become the supply side of tourism, must be designed and developed to meet the expectations of outsiders and residents alike. For example, tourists enjoy, as do residents, parks, scenic roadsides, entertainment, specialty food establishments, and streets that are clean, safe, and walkable at night. However, because the mindset and tenure of travelers are quite different from those of residents, communities and surrounding areas demand special planning and development for travelers that must also be compatible with the existing conditions.

A better term for "there" is *destination zone*. Tourist destinations do not originate on the marketer's desk but are the result of creative design and development schemes coupled with the appealing resources indigenous to a place. A destination zone can be modeled as illustrated in Figure 6-1. Its primary elements include *attraction complex(es)*, *focal community(-ies)*, *transportation and access*, and *linkage between the community(ies) and attraction(s)*. When destination zones are viewed in these terms, it is clear that rural areas are as much a part of the zone as is the community.

The process of discovery of potential destination zones within a nation or province is described in *Tourism Planning* (Gunn 1994). This is a process whereby the study of several factors within a region reveals areas where they occur in greatest quantity and quality. It begins with analyzing market demand and generalizing this information into two broad categories—activities that depend

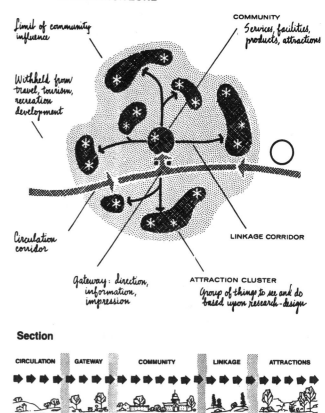

Figure 6-1. *Model of destination zone. Illustrated are the key elements of a destination zone: transportation and access (circulation corridor), focal community(-ies), attraction clusters, and linkage corridors. The cross-section illustrates tourist flow to and through these elements.*

58

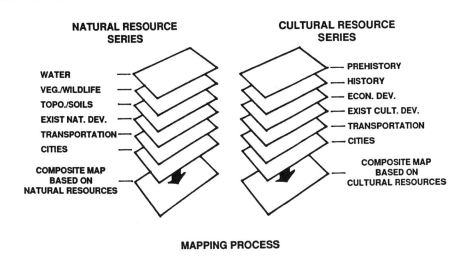

NATURAL RESOURCE SERIES

WATER
VEG./WILDLIFE
TOPO./SOILS
EXIST NAT. DEV.
TRANSPORTATION
CITIES
COMPOSITE MAP BASED ON NATURAL RESOURCES

CULTURAL RESOURCE SERIES

PREHISTORY
HISTORY
ECON. DEV.
EXIST CULT. DEV.
TRANSPORTATION
CITIES
COMPOSITE MAP BASED ON CULTURAL RESOURCES

MAPPING PROCESS

Figure 6-2. *Resource factors for regional analysis. Potential tourism destination zones can be identified by examining and mapping a series of foundation factors. Composite maps can then be prepared by means of computer Geographic Information System (GIS) programs.*

greatly on natural resource foundations and those that require cultural resources.

Then, the several resource factors are studied and mapped to reveal the areas that have the greatest strength. These resource factors are illustrated in Figure 6-2. After maps of these factors are prepared, they are overlaid on each other by using a Geographic Information System (GIS) computer process to create composites. The resulting areas can be generalized into zones, as illustrated in Figure 6-3 for the state of Illinois.

DESTINATION ZONE ELEMENTS

Tourism does not stop at the village or city boundary. Most of the planning and development problems described earlier in this book can be solved by reconsidering destinations as a whole. This procedure allows a broader perspective and stimulates greater development integration.

The "section" (cross-section) of Figure 6-1 dramatizes the need for integrated planning and development of destinations. The travel flow begins at the traveler's home and passes along water, land, or air circulation corridors. Every travel mode leads the traveler to a terminus, usually located in a community, where many attractions and services are located. The point where one enters a community from the circulation corridor is an important visual transition; first impressions influence the remaining experiences both negatively and positively. The traveler then moves on through linkage corridors to local and surrounding attraction clusters. Every one of these steps is a design and development challenge for developers and citizens alike. If some resources have not been developed into attraction clusters served by well designed linkage corridors and accessible community services, they may represent opportunities to create new destination

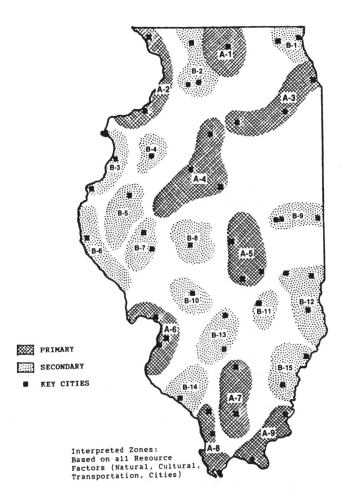

PRIMARY
SECONDARY
KEY CITIES

Interpreted Zones:
Based on all Resource
Factors (Natural, Cultural,
Transportation, Cities)

Figure 6-3. *Potential destination zones for Illinois. The regional analysis process was applied to Illinois, resulting in these potential destination zones. On the basis of research, recommendations for specific projects for each zone were offered to tourism leaders.*

zones. It must be emphasized, however, that destination zones are based as much on market forces as on intrinsic physical resources.

Following is a brief description of the major elements of destination zones.

Attraction Complexes

Attraction clusters form the backbone of tourist destinations. After tourism leaders and citizens thoroughly study the resource potential of an area, there may be opportunity to develop new attraction clusters that are of interest to several market segments.

Indeed, the future success of an attraction cluster, in terms of visitor satisfaction, business success, and resource protection, rests heavily on planning and development decisions. Several scenarios are possible. The development may be so inept that the term "tourist attraction" comes to mean "tourist trap." Or the design may be so ordinary that the attraction fails to sustain tourists' interest. Conversely, the attraction cluster can be planned to make outstanding use of resources and enrich visitors while supporting profitable business.

An example of an attraction cluster is illustrated in Figure 6-4. On the basis of the cluster design principle, it

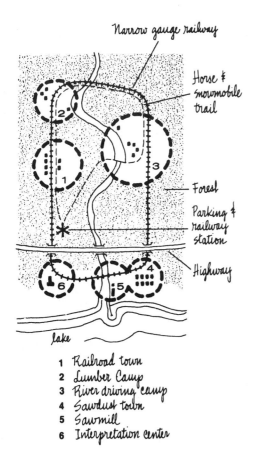

Narrow gauge railway

Horse & snowmobile trail

Forest

Parking & railway station

Highway

lake

1 Railroad town
2 Lumber Camp
3 River driving camp
4 Sawdust town
5 Sawmill
6 Interpretation center

Figure 6-4. *Concept for an attraction complex. This complex, based on significant historic events, was recommended as part of a study of and plan for Michigan's Upper Peninsula. Sketches illustrate potential elements of this complex.*

National parks, historic sites and preserves often serve as attraction complexes with a variety of features of interest to visitors—scenery, historic buildings, trails, water resources, and other outdoor recreation opportunities. (Photo Buffalo National River: Kenneth L. Smith, courtesy National Park Service)

can be concluded that a dramatic, interesting, and educational attraction complex could be developed in the Great Lakes timber region of the United States. From 1880 through 1930, the harvest of the timber used to build the frontier cities of the Midwest was the foundation of this region's economy. Today, this chapter in American history is nearly forgotten, but research shows that several aspects of the area would lend themselves well to redevelopment. By incorporation of a river, a harbor, and related sites that played important roles in history, a fascinating complex could be restored, with the following six attractions:

Railroad town

Features could include a restored version of the community, with a general store, cabins, a steam sawmill, a blacksmith shop, and bars. Exhibits could include equipment, machinery, and processes of the era.

Lumber camp

Features could include a restored cabin with a smoke hole in the roof and an open fire; bunkhouses; a cook's cabin; a beanhole (a pit in which beans are cooked); barns;

a granary; a blacksmith's shop; logging trails; skidding trails; skidways; sleigh trails to river landings; and banked logs at landings. Exhibits of tools and equipment used in the prerailroad era and action displays of axe cutting, cross-cut sawing, marking, skidding, sleighing, and banking could be provided.

River driving camp

This camp could consist of tents, an outdoor cooking area, dams, piers, and a boom-company farm (a company that manages log booms). Exhibits could include oxen teams, bateaux, peaveys, poles, a cook's outfit, and a blacksmith's outfit. Action displays could include using a bateau, breaking a log jam, eating while balancing on a log, and birling.

Sawdust town

The possibility exists of restoring actual buildings on their original sites. Features could include bars, restaurants, and a floating night club, all of which could simulate the buildings' original interior design, exterior design, and functions. The flavor, glamour, zest, and character of the lumber era would dominate. Here also

For travel destinations, communities provide essential tourism functions but are best located outside major attractions. Shown here is Gatlinburg, Tennessee, a gateway community for Great Smoky Mountains National Park. (Photo by Ben Humphreys, courtesy Gatlinburg Chamber of Commerce)

could be located ample display and sales areas for souvenirs and gifts. This grouping could offer some of the commercial and more frivolous establishment, such as saloons, that would be out of place in other parts of the complex.

Sawmill

Features could include a restored pre-1900 sawmill; a pond; river storage; booms, piles, log pockets, river tugs, and lumber schooners; catchmarking sheds; tally shanties and scaling gaps; boarding houses; a refuse burner; and a sorting shed. Action displays could reenact the complete process of producing lumber in the original manner.

Interpretation center

Sponsored by a modern timber company or educational institution, an interpretation center could feature indoor and outdoor exhibits depicting the history, science, and modern technology of forestry and forest products. Action displays could include working models with taped narrations.

The entire complex could have an entrance and exit control point easily accessible from main transportation routes. Rather than build drives and parking lots to bring visitors to all of the features, designers could restore a turn-of-the-century narrow-gauge railway for access and to help create a stimulating atmosphere. The redesign of both land and building features would require skill and creativity together with research results from land analysis. Planners, local citizens, and specialists in logging history could produce an enjoyable and meaningful travel experience centered on a significant chapter of America's past.

Service Community

As discussed above, a service community is an essential element of any destination zone, and this fundamental must be enhanced through better planning. When all public and private development policies consider the visitors' needs alongside those of residents, destinations will be able to fulfill their functions more efficiently and beautifully.

The support community offers many tourism functions. It is often the traveler's main objective, especially for shopping, entertainment, visiting friends and relatives, and business and professional activities. It is the primary location for lodging, food services, civic amenities, traveler assistance, and transportation termini. Its appearance, cleanliness, and safety are paramount.

Because communities originated from economic and social reasons other than tourism, their conversion to tourism is a major challenge. As was described in

Chapter 1, communities have the opportunity to guide tourism development in ways most beneficial and acceptable to them. Land use issues are critical to this process. Policies are needed to guide new tourism development where it can be most successful and yet retain the basic community values that are important to residents.

Frequently, a regional analysis reveals the opportunity for several communities to function as service centers of a tourism destination. Communities that have been rivals in the past may become partners in a new tourism enterprise based on a common destination theme centered on shared natural or cultural resources.

Community development for tourism demands great care in planning physical development and programs.

Transportation and Access

The design of key circulation corridors and their accessibility to the community deserve special consideration. These corridors, predominantly highways, must be more than just pipelines to such services as hotels, eating places, entertainment, and stores for tourists traveling to attraction clusters. Their design must go beyond structural engineering and include greater consideration of travelers' functional needs and visual impressions along the way. Better signage for directions and information is needed. Cooperative agreements and controls are necessary to prevent roadside blight and enhance road-

Today's use of logos on signs and computer information systems eliminates the need for excessive intrusions of billboards and signs on scenic vistas. (Photo courtesy Coalition for Scenic Beauty)

side vistas. Better rest stops require locations near service centers rather than in isolated areas where travelers feel less secure. Travelers prefer stops that have information, telephones, snacks, meals, soft drinks, maps, gasoline, rest rooms, and other products and services.

Linkage Corridors

Linkage corridors tie attraction clusters, the vital organs of destinations, to community services. These corridors often slice through the worst of communities' slums and industrial areas. Traditional engineering emphasis on road design for local traffic is insufficient for visitor traffic. Much greater sensitivity is needed regarding travelers' perception, needs, and typical functions. These corridors are entranceways to attractions, and so all roadside views take on great importance in setting the mood for travel objectives. Even roads that pass through ugly business strips or ramshackle neighborhoods can be enhanced greatly through new landscape controls and actions of redesign, regrading, and relandscaping.

Together, the interrelated and interdependent functions of these four main elements in a community and its surrounding area make up the destination body. When the concept of a new destination is implemented, many jurisdictions, designers, developers, managers, and local citizens must bring their individual ideas to a common table. The many stakeholders in and visitors to the destination will benefit as a result.

MARKET USE OF ZONES

As described in Chapters 3 and 4, tourism development can be generalized into two broad categories—touring circuits and longer-stay. However, these categories often overlap. Former resorts today cater to visitors passing through. Present resorts now add tours of surrounding areas to their lists of offerings. Cruises that are predominantly touring often include stays at port cities and surrounding areas. Convention and conference centers often add pre- and postconference tours to their programs.

From a community and destination perspective, greater economic potential may be realized when both touring-circuit and longer-stay tourism are provided in one destination zone. To accomplish this objective it is

Once blighted by billboards, this view of scenic Hawaii across the Ala Wai Canal, Waikiki, is now possible because of citizen action by the Outdoor Circle of Hawaii.

often necessary to make changes in supply-side development.

For example, the touring-circuit mode, stimulated by policy and promotion, has been the dominant form of tourism in the Canadian province of Nova Scotia. The province's *Travel Guide* (Tourism Nova Scotia 1990) contains descriptive information (attractions, accommodations, campgrounds) and travel maps organized into nine geographic area tours. For visitors traveling by personal car or motorcoach, this guide provides excellent help for full enjoyment of the region.

However, recently the communities of the province have raised questions about this exclusive emphasis and the lack of concern over their longer-stay potential. New support for community-oriented destination development is expressed in the provincial report *A Strategy for Tourism in Nova Scotia* (Tourism Nova Scotia 1995).

For example, the community of Yarmouth, the key entry port for U.S. travel markets, has recently laid plans for increasing its attractions for longer-stay visits. In the past, because it is the entry point for two trails, the Evangeline Trail and the Lighthouse Route, its tourism was divided and focused only on pass-through traffic. In 1988, the Yarmouth County Tourist Association and Yarmouth Development Corporation established a task force to study and redesign the area as a longer-stay destination (Deveau 1995). A focal point is to be the Cape Forchu Lighthouse and its site. Other potential attractions include cycling, hiking, and canoe trails; museums; manufacturing and processing plants; scenic routes; performing arts and other events and festivals.

Throughout the world, new market trends toward adventure, nature, and culture tourism are stimulating rural areas and small towns to consider new destination development. Because of their small size and past traditions, it is especially important for such communities to plan carefully so that they avoid negative impacts, which can be more damaging to them than to larger cities.

URBAN–REMOTE SCALE

Figure 6-5 illustrates the destination zone concept in another way, by applying an urban-to-remote scale. For convenience, four subzones are suggested. This model brings even remote parks into the context of the overall destination complex and further reinforces the four essential elements of destination zones described above.

The *urban subzone* is the prime location for motels, hotels, restaurants, bars, and other tourist services as well as major attractions. Arenas for pageants, cultural events, and festivals are of great importance to the visitor to urban areas. Most historic sites and buildings are to be found in urban areas, and many have waterfront services and water-oriented activities. Cities are the beginning points for scenic tours and the subzone in which local parks, zoos, swimming pools, and spectator

Figure 6-5. *Urban–remote scale. Another approach to destination planning involves analyzing the differences in development potential from urban areas to remote locations.*

sports arenas are located. The urban setting demands special land use designs and controls for the best utilization of its resource base and adaptation to its residents' needs. Especially important are concentrated resort communities, where the large volume of visitors in some seasons requires special overall design. For example, the great influx of Northerners to Southern resorts in winter puts great stress on streets, parking, shopping, parks, and police control, originally designed for small local populations.

The *suburban subzone* is the place where bowling alleys, general indoor recreation facilities, amusement parks, and night clubs tend to be found. Here is also the potential for industrial tours, which are attractions of increasing importance. Suburbs are also good locations for overnight campers who do not seek the amenities of outstanding natural resources. Finally, they are the dominant areas in which travelers visit friends and relatives.

In the *rural subzone,* attractions based on natural resources seem to dominate, whereas service businesses thrive better in urban or suburban subzones. Here, natural resource-oriented camping, farm-oriented recreation, ranch resorts, snowmobiling, and water-based recreation are found, along with well designed vacation homes and all types of water-oriented resorts. Lakes and reservoirs in this subzone are within easy reach of community centers that draw visitors from outside as well as within the destination zone. In hilly northern regions this type of subzone is well suited to winter sports development.

The *remote subzone,* farthest from the urban center, has site and location advantages for a variety of attractions and tourism uses. Generally, these do not include masses of facilities and services such as hotels and restaurants, which can function more efficiently in the urban area. The remote subzone is better suited to national parks and preserves, where primitive camping and other outdoor activities are appropriate, such as trail riding, hiking, nature study, hunting, and fishing. Rural areas and small towns in this subzone are increasingly seeking to derive economic benefits from tourism by establishing tourist services. As is emphasized throughout this book, this subzone contains the greatest volume of resources needing protection, requiring especially careful planning. Issues of competition from timber harvesting, mining, and other extractive industries must be dealt with.

Shown in Figures 6-6 and 6-7 are a sketch plan and an aerial perspective of a hypothetical small destination zone. These figures emphasize the importance of attraction complexes and their relation to the other functional elements of a destination.

The scenario in Figure 6-6 begins with access from a major highway. At the interchange, a specially designed landscape corridor captures the regional theme and avoids the usual visual pollution of billboards and junk. Directional signs have been replaced with a complete information plaza situated near the interchange and also easily accessible from the airport.

The community provides urban attractions and all basic travel services, including lodging, food, entertainment, shops, car service, and other travel needs.

Radiating from the city are attractively designed travelways leading to attractions suited both to a variety of markets and to the resources of the destination area. Illustrated here are sixteen potential attraction complexes.

Figure 6-7 presents an aerial view of the plan. The four areas described in Figure 6-5 are keyed to this view. The basic theme of this concept is the relationship between people and land. The design of destination zones must reflect the travelers' purposes as well as the location's environmental foundations.

FALLACIES AND HALF-TRUTHS

A review of approaches to destination development in several countries reveals a variety of characteristics important to planners and developers.

Perusal of only one report on destination zones may give the impression that the definition of "destination zone" is uniform. This is not the case. Some documents identify only administrative zones cutting a region into segments suited to managing tourism programs, whereas others identify market zones suited to certain marketing strategies. Some consider only a specific site development, such as a resort complex, as a destination zone. Recently, as described throughout this book, others have identified potential destination zone determinants on the

basis of research and analysis of contributing geographic factors.

It is a fallacy to consider destination zones as having definite boundaries. This erroneous belief is supported by boundaries on maps. The edges of destination zones are not sharply defined and they should be depicted as bordered by only transitional belts. Although the areas represented within boundaries have some cohesive properties, they defy political definition and legal description.

Another danger of zone graphics in reports is the implication that they are permanent. This belief may be correct in rare instances, but it is much more likely that future circumstances will alter them. Changes in access, resource quality, political policy, international currency exchange, market preferences, visitors' ability to pay, and supply can alter the shapes and sizes of destination zones.

Some communities and governments believe that destination zones can be willed into reality. Although enthusiastic local commitment and government support can assist development, such factors as the geographic setting, resources, access, and a viable service center must also be present.

Even in a market economy, destination zones are not the sole prerogative of the private sector. Most governments are heavily involved in destination development and promotion. They often own the natural resources that make an attraction viable, provide policies and incentives, and develop infrastructure such as streets, lighting, water and waste systems, and police and fire protection.

Whereas intrinsic physical properties are important to destination zones, they are not viewed with equal interest by all visitors at any one time or over time. Resources such as water, forests, and wildlife are basic to many attractions across diverse cultures. However, government policy, economic changes, and access and market shifts can elevate or lower the value of such resources as destination factors.

Frequently, destination zones are thought of only as tourist activity areas. This concept omits the essential support of their service communities. Owners of isolated resorts have learned that without the support of a service community within reasonable range, their success is impaired because of poor access and the high costs of development.

There was a time when environmentalists believed that a mere inventory of resources determined a destination zone. This narrow focus on the environment left out essential considerations of travelers' objectives and the capability of an area to support diverse development. Inventorying is nondirected; only by serendipity does it produce something useful. On the other hand, a purposeful analysis of resources can lead to the identification of areas that have the resource base to become potential zones that meet market demand.

That destination zones can be scientifically determined is another half-truth. A factual understanding of both resources and markets, achieved through scientific methods, is essential. Certainly all developers should take advantage of all facts. However, the capriciousness of markets and the creativity of designers and developers are equally important determinants, albeit less scientific.

The identification of destination zone potential cannot ensure the feasibility of all enterprises. Regional studies and analyses can assist the developer. However, true feasibility can be determined only when individual site developers and designers agree on specific market and financial details.

Finally, community leaders and developers can realize valuable tourism potential by guiding its growth, not just within a community but within the overall destination. As described by Mathieson and Wall (1982), when an examination of key factors within a destination area is made, a sound foundation for future development is established. These factors include natural resources, economic resources, local cultural and social characteristics, politics, and the level of existing tourism development.

CONCLUSIONS

The main conclusion from this discussion of tourism destinations is that the diversity and abundance of site development must always be seen in a larger context. The development of individual sites for services, attrac-

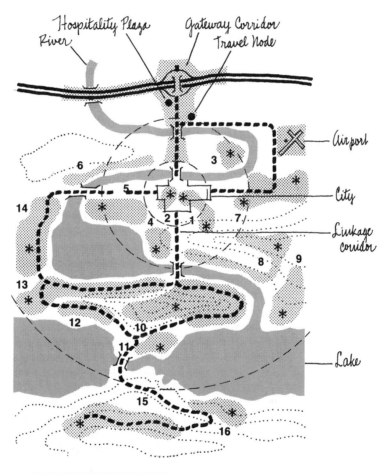

ATTRACTION COMPLEXES

1 Historic restorations
2 Entertainment, meeting
3 Urban camping
4 Resort hotels
5 Golf, picnic, marina
6 Scenic drive, cruise
7 Industrial tours
8 Scenic drive, overlook
9 Vacation farms
10 Low density vacation homes
11 Resort camping
12 Organization camps
13 Medium density vacation homes
14 High density vacation homes
15 Wildland camping, hiking
16 Wildland resort

Figure 6-6. *Concept of destination zone. A hypothetical destination zone indicating elements of potential and relationship to the urban–remote scale.*

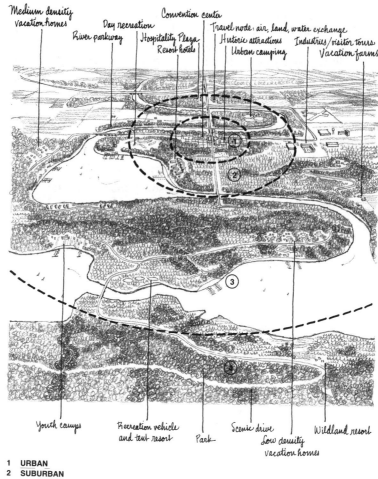

Medium density vacation homes
Day recreation
River parkway
Convention center
Hospitality Plaza
Resort hotels
Travel node: air, land, water exchange
Historic attractions
Urban camping
Industries/visitor tours
Vacation farms

Youth camps
Recreation vehicle and tent resort
Park
Scenic drive
Low density vacation homes
Wildland resort

1 URBAN
2 SUBURBAN
3 RURAL
4 REMOTE

Figure 6-7. *Aerial perspective of destination zone. A sketch illustrating the same features shown in Figure 6-6 and the four zones shown in Figure 6-5.*

tions, and transportation will continue. However, every site is connected to and dependent on many others, even across larger geographical areas.

The several conceptual models presented here demonstrate key functional relationships. If businesses are to succeed, tourist experiences to be richer, and resources to be protected effectively, planners and developers must be more sensitive to all these functional relationships.

Those who seek community tourism development will be hampered severely if their programs stop at their city boundaries. Their success depends as much on the surrounding area as on the areas within those boundaries. Both visitor rewards and the tourism economy are enhanced when the destination zone becomes the unit for planning and development.

The focal point for the destination zone is one community, or several. This is the terminal point for all market access; the community contains many amenities, services and infrastructure essential to tourism; and it is here that governance and leadership can be found.

Equally significant to development, and more difficult to integrate, is the surrounding land area. The many tourist activities associated with outdoor recreation and natural resources are usually located here. These lands are usually owned by residents and governed by officials outside the communities.

To fully understand and take advantage of the relationship between communities and outer lands usually demands entirely new education, new cooperation, and perhaps even new organization and legislation. When the residents within this larger unit, the destination zone, begin to understand their collective potential for tourism, they can initiate new processes and mechanisms for planning their future. Based on this initiative they, not outside investors or governments, can establish the blueprint for guiding tourism growth in a manner that retains the integrity of their community.

Spatial Patterns

As an aid to the development of tourism, this chapter describes geographic relationships for the several attraction classes described in Chapter 5 (Table 5-1). These relationships are modeled in several diagrams to show how access, communities, lodging and food services, attraction clusters, and linkage are integrated. Because several classes of attractions are similar in function, they may be grouped under a single model. Following is a list of models and attraction classes for touring-circuit and longer-stay attraction development (Table 7-1). Although not exhaustive, this list shows the classes and models that represent the majority of tourism development patterns today.

TOURING CIRCUITS

The great diversity of market push and resource pull today stimulates the development of a great variety of places for touring. The following four models diagram the functional relationships for thirteen classes of touring activities. They should be useful in guiding planners and developers toward creating better functional relationships at these destinations, especially linkage with communities.

The diagram in Figure 7-1 (Model A) illustrates functional relationships that are similar for several travel purposes: visiting roadside scenic areas, outstanding natural areas, and touring circuit camping areas.

This diagram raises several design issues. First is the question of coordinating the design, management, and control of scenic tours. Many firms and agencies produce maps delineating such tours. Others design and manage roads. Still others control roadside vistas, natural areas, and campsites. Some catalyst is needed to bring the several actors together for integrated function.

Second is the question of theming. Because of market segmentation each touring circuit requires a dominant theme in both its design and promotion.

A third concern is the length of tours. If a tour is to be created, its length must be in accord with travelers' time constraints. One solution is to divide long tours into several independent segments. Too much travel between destinations is undesirable.

Fourth, it may be necessary to legislate tour design, at least for some scenic drives, as has been done in Wisconsin (Wisconsin Rustic Roads Board, n.d.). Zoning laws and other regulations of land use may be the only

TABLE 7-1. List of Model Diagrams

Touring circuits		Longer-stay	
Model A:	Roadside scenic areas	Model E:	Resorts
	Outstanding natural areas	Model F:	Longer-stay camping areas
	Touring camping areas		Hunting/water sports areas
Model B:	Water tour places	Model G:	Organization camping
Model C:	Homes of friends and relatives	Model H:	Vacation home sites
Model D:	Unusual institutions	Model I:	Festival and event places
	Shrines and cultural places	Model J:	Convention and meeting places
	Food and entertainment places		Gaming centers
	Historic buildings and sites		Sports arenas
	Ethnic areas		Trade centers
	Shopping areas		Science and technology centers
	Crafts and lore places		
	Air tours		Theme parks

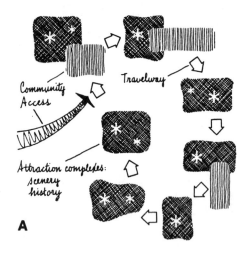

Figure 7-1. *Model A. A functional diagram showing spatial relationships between attraction complexes, communities, and access and traveling routes. Similar relationships apply to roadside scenic areas, outstanding natural areas, and touring camping areas.*

ways to protect certain unusual scenic aspects of roadsides. In the American culture, appropriate roadside views for touring include such land uses as forests, wildlands, historic sites, and land uses such as agriculture. Excluded from designated scenic roads should be landfills, major excavations, heavy industry, signs, billboards, and derelict structures.

Figure 7-2 illustrates a model "hospitality plaza," a concept meant in part to eliminate billboards, a major cause of scenic degradation. By pulling off the highway for information and other travel services, the traveler would no longer be dependent on highway signage. Published guides, maps, directories, interactive information kiosks, and general travel literature would be available. In more heavily traveled areas, trained counselors

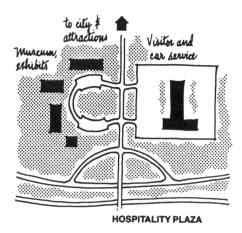

HOSPITALITY PLAZA

Figure 7-2. *Hospitality plaza. A freeway exit is a prime location for a visitor hospitality plaza. Included should be visitor information, a museum, exhibits, car services, and a lunchroom to provide an introduction to the area.*

could offer information about attractions, services, and facilities. Computers could be used to reduce manpower costs and provide a wealth of guidance to the traveler. Up-to-the-minute information about weather, highway conditions, hours of admission, fees, and capacities of certain attractions could be incorporated into such a system.

The plaza concept could also include a major interpretation function. Indoor and outdoor exhibits could illustrate important points of interest along routes. Models, samples of products, and dioramas could induce the traveler to investigate attractions nearby.

To serve such functions, information plazas would have to be available throughout a nation. In fact, plazas owned by private enterprise and strategically located throughout a country could completely eliminate the need for roadside promotional signage.

Figure 7-1 (Model A) offers a functional relationship for three kinds of attraction classes.

Roadside Scenic Areas

Observing roadside scenery by motor coach or personal car continues to be an important travel interest. But terms such as "shunpike" and "blue highway"—reactions to the deterioration of freeway roadside scenery—suggest real challenges for highway design today. Better highway alignment and better scenic sensitivity should be added to the traditional engineering criteria for structural design. It is unfortunate that the ethic of roadside beauty is so lacking that specially designated scenic highways must be legislated.

Outstanding Natural Areas

People often take tours primarily to visit especially interesting natural areas along the way to a destination. National, state, and provincial parks and natural reserves contain an abundance of flora and fauna of interest to the nature travel markets.

With regard to natural areas, two design issues are particularly important. First, designers need to analyze the travel corridor from the main route to a natural area for its aesthetic appeal, and to remove anything that detracts from the beauty of the setting. Second, they must address the issue of traveler services. If food, educational, and informational services are needed, they can be incorporated into the corridor in ways that complement the final tour objective.

Touring-Circuit Camping Areas

Camping continues to be an important traveler activity. While on tour, this market segment prefers locations near urban areas rather than special scenic areas. Campers who stop at a place for only one night require car services, food, occasionally motels, and access to the amenities of nearby cities. Planners and developers should accommodate local camping trends, such as increased RV travel. Many camp sites, especially those in warm climates, need facilities to serve both touring-circuit travelers and those using the area as a winter destination.

Water Tour Sites

Figure 7-3 (Model B) illustrates an increasingly important travel purpose—cruises. The diagram illustrates the functional relationships between the major attraction complexes, community service centers, and access.

Today, travelers tour by boat for business as well as pleasure. Conferences and seminars take place on board personal cruisers and huge passenger liners as well as at urban convention centers. Water travel, no longer a

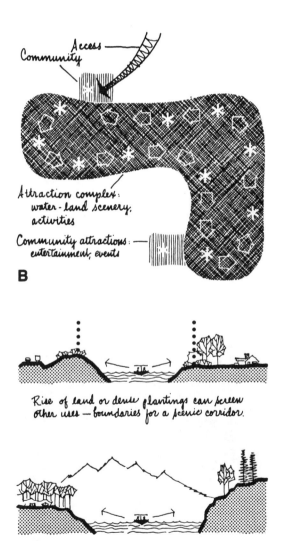

B

Rise of land or dense plantings can screen other uses — boundaries for a scenic corridor.

Water-to-land vistas generally upward — produces feelings of protection, enclosure.

Figure 7-3. *Model B. Functional relationships for open water and river tours. The linkage between water and land offers special development opportunities.*

transportation mode but rather a destination complex itself, requires special design consideration.

Ports and harbors, usually located at the "back" sides of cities, need new design treatment to accommodate tourist visits. These entrance points are gateways to exotic cultures and require informational and transitional service centers.

Waterway environments—streams, rivers, canals—offer a unique tourism experience that is gaining in popularity worldwide. The most outstanding characteristic of inland waterway settings is enclosure. These water courses are always bounded by shorelines, which present a visual edge. As this edge changes with the progress of the boat, so do the views. Even the smallest land promontory becomes a major curtain, withholding disclosure for a moment and then revealing a fresh panorama. Coastal waterway environments usually offer broader vistas than terrestrial environments.

Variegated land use patterns along the shore add interest to a water cruise. But, unfortunately, the sequence is sometimes broken by such intrusions as waste dumps, eroded banks, dilapidated buildings, or, worst of all, industrial outlets pouring forth odorous and toxic wastes.

When these hazards are eliminated, waterscapes offer a unique recreational experience. The water edges of farms, forests, parks, and some cityscapes constitute a separate world of vistas. Objectionable inland land uses can be entirely screened out with only a little landscape development along the edge. The water corridor is an inversion of most highway corridors. Overlooks are turned into "upwardlooks" (except along rivers enclosed by levees). Long vistas are over the water, not the land. Furthermore, boat cruising allows much greater maneuverability than land travel; one can dawdle, stop, or cover great distances in a short time.

The key to the planning for a water-based environment is the land. The relationship between water and land produces the interest, the tranquility, the drama, the excitement, and the fun of water touring. For example, the reflection of buildings and lights in the night waterscape along a community is magnetic and unique. It can be a spectacular attraction; and, indeed, the waterscape is designed more as an attraction than as a travel corridor, although it serves both purposes. A shoreland scenic drive concept is shown in Figure 7-4. Locations for the concentrations of lodging, food, and other services in the community are suggested, with the surrounding resource-based attractions under protective control. By designing special sites for visitor use—beach, overlook, historic site, trail, scenic auto tour—planners can offer visitors enrichment and enjoyment without damage to the resources beyond.

To avoid conflict, high-volume visitor uses of water may need to be allocated on a temporal or spatial basis. Some waters are unique for their scenic appeals, others for their water sports potential.

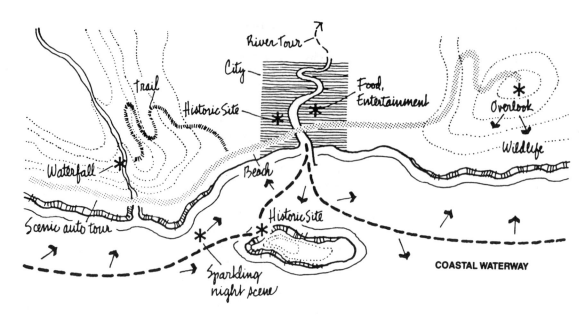

Figure 7-4. *Coastal waterway. When this area is designed for both water and land touring, a rich and interesting experience is available to several travel markets.*

Homes of Friends and Relatives

Visiting friends and relatives is an often-forgotten but dominant travel purpose, illustrated in Figure 7-5 (Model C). This function goes beyond mere contact with people within their residences. Many attractions are visited by travelers because they are en route to or near the homes of friends or relatives. Frequently, a visit gives the host the opportunity, and sometimes the challenge, to provide a variety of interesting things to see and do in and around his community. The host residence is used as a base for circuit touring to parks, museums, scenic areas, sports arenas, entertainment, and specialty food services.

Do traffic engineers and city planners consider design linkage for visitors when they establish street systems for residents' homes? Many travelers can testify that no such overt design policy exists, except in rare instances. This function is so important for all travel that it should be given higher priority than it receives today.

Many tours have as their objective urban destinations, illustrated in the diagram in Figure 7-6 (Model D). Although their specific purposes may be different, the same general functional relationships apply to several travel objectives for both residents and visitors in cities.

Pedestrianism is essential to the design of community tourism. For half a century, urban designers paid increasing attention to the automobile, but now a reversal—a renascence of the city walker—is finally under way. True, traffic problems must be solved, but solutions can benefit rather than alienate the pedestrian. Strictly one-sided, heavy-handed engineering schemes, even those fostered by city ordinances and policies, have turned

many cities into asphalt and concrete deserts and pushed the pedestrian into unsafe, inconvenient, and hostile environments. Tourists must get out of their cars to enjoy a community, as must local citizens, of course. A comprehensive new pedestrian-oriented environmental design concept is required as recommended by Wiedenhoeft (1985). He suggests that in addition to stimulating downtown's function as a regional magnet, this new thinking will enhance uniqueness and appeal of place, will reinforce the residential function of cities (including downtowns), will solve traffic and pollution problems,

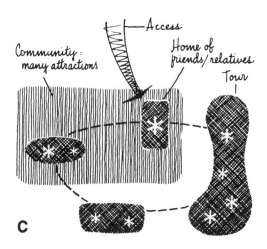

Figure 7-5. *Model C. The popular travel activity of visiting friends and relatives deserves special planning and development. Usually incorporated into visits are tours and events in the community and its surrounding areas. Urban planners should include these visitor functions within their scope of responsibility.*

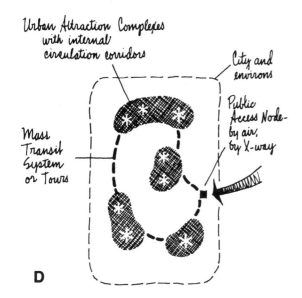

Figure 7-6. *Model D. Cities are frequent travel objectives. Planning consideration should be given to all elements, such as institutions, shrines, cultural and historic sites, ethnic areas, shopping centers, travel services, craft outlets, and entertainment.*

and will contribute to furthering a whole range of social and cultural goals.

Following are eight classes of touring circuit attractions illustrated by the functional diagram, Model D.

Institutions

Frequently forgotten in tourism planning is the role of such places as manufacturing plants, universities, religious complexes, and medical centers, which can be developed for specific travel segments. Hotels near medical centers now recognize the importance of catering to outpatients as well as hospital visitors. This clientele often needs special architectural and site design as well as visitor-oriented programs.

Travelers have discovered the educational value—especially for children—of visiting manufacturing and processing plants. Special safety protection and interpretation systems are required if the public is to be allowed access to such places.

College towns often lack adequate ancillary travel services for their visitors. Special activities for spouses of those attending seminars and conferences, proper transportation, and special interpretation of local resources should be woven into the planning fabric of these communities.

Shrines and Cultural Sites

Important national, provincial, or state cultural places, such as monuments, shrines, churches, and museums,

which are attractive to many touring-circuit travelers, are usually located in urban settings. Because these places were not originally designed for mass use by tourists, many questions must be addressed if chaos is to be avoided. Do the site and building features need to be modified? How can conflict between prime users and tourists be avoided? (Such conflict has been a problem, for example, when tourists have disturbed church services in major historic cathedrals with excessive noise, camcorders, and flash photos.) Are there adequate toilet facilities and parking space for visitors? Can these be so designed as to retain the aesthetic qualities of the place? These and many other design and management problems can be solved, but only if owners and developers recognize the need to do so.

Food and Entertainment Sites

Because so much of food service is local, its several design and development implications for travelers are often overlooked. Access to lodging and attractions is important to travelers and frequently can be incorporated without compromise into location, site, and building designs for local trade. The design problems for entertainment places center on how conveniently visitors can find them by using taxis, mass transit, or their own cars. Planners and tourist information specialists need to collaborate in helping visitors find entertainment places.

Historic Buildings and Sites

The recent surge of interest in historic places has stimulated a great amount of redevelopment, both good and bad. Most frequently, agencies and organizations that redevelop historic places do so to preserve them. This is a worthwhile objective, but visitors must be considered as well and provided with access, parking, and creature comforts. Efforts of designers and programmers to link several sites together may offer more meaningful presentations of history and culture.

As was discussed in Chapter 5, the principle of authenticity must be observed in historic redevelopment. In some areas, well-meaning promoters and developers have overdone "flea market" architecture. Whereas a colorful decor that evokes the past has appeal, especially when it is indigenous to a community, imported schlock may produce a landscape as homogenized as the design it replaced. Special land use and design regulations may be needed to protect historic districts from modern intrusions.

Ethnic Areas

Converting ethnic areas into use for visitors is a sensitive design, development, and management problem. The desire to experience a different culture is one motivation for travel, but it is difficult to insert masses of strangers into a cultural setting without social conflict and physical damage.

Any designed development based on ethnic customs, artifacts, products, or structures must be supported by the local ethnic group in question. Outside investors may cause strong social conflict by developing attractions that make spectacles or freaks out of local populations. However, when ethnic groups wish to share their past culture and distinctive ways of life, attractions can be rich and rewarding for the visitor. Lancaster, Pennsylvania, has solved this problem by allowing a commercial zone for tourist contact with its Amish culture and prohibiting visitors from invading the privacy of the Amish in the rest of the community (Buck and Alleman 1979).

Shopping Areas

For some travelers, shopping is the greatest travel objective. Too often in communities, only resident markets are considered when sites are selected for shopping centers. But travelers as well as residents wish to buy clothing, film, drugs, camping supplies, souvenirs, and food items. Locations that consider visitor as well as local markets can enhance sales. Feasibility consultants, developers, and designers of shopping complexes should modify their policies to incorporate this consideration.

Craft and Lore Sites

Travelers en route to a destination are attracted to sites offering special crafts, paintings, tapestries, carvings, and sculpture, as well as to places that are known through legends and lore. The development of such sites, especially when they are located near other attractions, can add materially to the pulling power and visitor interest that derive from the larger attraction complexes.

Air Tours

In some regions, the construction of roads or trails would be damaging to the resource base, and sightseeing is best accomplished by helicopter. Well-designed heliports should be associated with other access routes for travelers, as well as with food and lodging services. However, air tours, as has been found in popular Grand Canyon National Park, can create severe safety hazards and noise and air pollution.

One of the greatest advantages of air travel is its geographic scope. Yet this asset is overlooked by all but the most avid landscape buff. Generally, the air traveler has very little help in relating himself to the land he is traversing.

Many travelers on commercial airline flights would probably appreciate taped narrations of their entire flights from take-off to landing. Airlines would not have to produce many tapes, because air routes and travel times vary only slightly from flight to flight; such a project would be economically feasible. The narrations could describe cities, farmlands, mountains, and other points of interest en route. Explanation of the geological formations, historic background, present industrial development, and recreational opportunities of these places might stimulate passengers to visit them on subsequent trips. Maps and descriptive folders, keyed to the narration, could heighten interest.

LONGER-STAY TRAVEL

In recent years air and expressway travel has enabled tourists to spend longer periods of time at their destinations. When people travel long distances, they tend to have less interest in attractions en route than in those at the destinations themselves, where travel objectives are usually clustered in one general vicinity.

In the following discussion twelve classes of longer-stay travel purposes are illustrated by six functional models. In addition, the special cases of national parks and coastal tourism are described.

Resorts

Figure 7-7 (Model E) illustrates the general relationship between community, access, attractions, and services for resorts. Conceptually, the modern resort is similar to the organization camp in its relative independence and tight linkage with attraction clusters. Key planning considerations are the physical and aesthetic development of attraction clusters, linkage with housing and food service, and linkage with the community. Compatible neighboring land uses include other resorts, second-home subdivi-

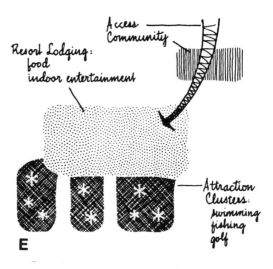

Figure 7-7. *Model E. Although the linkage between communities and longer-stay attractions may not be as strong as with touring, it remains an important planning function.*

sions, and extensive-use public parks. Certain natural resources, such as a warm climate, may encourage the creation of new resort cities. In such cases, designers and planners must encourage the protection of basic attraction resources.

In the United States and Canada, ranch resorts with cowboy and western themes continue to gain in popularity. These vary along a scale from operating farms and ranches to contrived complexes in ranch-like settings. Farm and ranch businesses maintain their primary functions, perhaps with only slight remodeling of the owners' homes to accommodate visitors, allowing merely a few guests at selected seasons. At the other end of the scale are huge complexes with lodging, food service, country and western entertainment, and other recreational amenities, such as golf courses, tennis courts, pools, and equestrian trails. Such complexes require new design input.

Camping Areas

Figure 7-8 (Model F) illustrates two classes of development—natural resources camping areas and hunting and water sports areas.

Adults and families continue to use RVs to get to these destinations. Older and retired RV users who spend the winter in warm climates usually prefer community amenities nearby. Park sites often contain central activity rooms, golf courses, hiking trails, and bicycle trails.

Frequently, package tours of these sites expand opportunities for enjoyable and healthful activities. In northern climates, summer use of natural resource attractions continues to be popular for all age groups. Those who design such places must try to provide attractive settings for specialized market segments that stay at a destination for an extended period of time.

Hunting and Water Sports Areas

Those who engage in hunting and fishing expeditions may camp near a resource area or occupy commercial lodging in the nearest community. All facilities close to or within the resource area should be designed to be compatible with the special resource features and to emphasize environmental sustainability.

Organization Camping Sites

For many years young people and adults have engaged in forms of camping exclusive to each group. The functional relationships relevant to this purpose are shown in Figure 7-9 (Model G). Examples include camps led by the Boy Scouts, the Girl Scouts, Young Men's Christian Associations, Young Women's Christian Associations, and 4-H clubs, as well as by church and conservation groups. In addition to the main goal of living close to nature, such groups may have educational, religious, or fraternal objec-

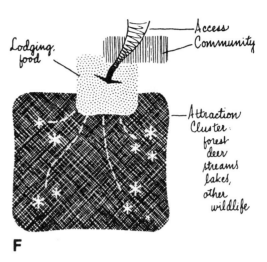

Figure 7-8. *Model F. Functional relationships are quite similar for longer-stay camping areas and hunting and water sports areas. Natural resource settings dominate the attractions.*

tives. The development of organization camps is promoted heavily by exponents of outdoor and conservation education, now known as "environmental education."

Although it is a key destination activity, organization camping land use is conceptually the reverse of many others. Attraction clusters surrounding a core of facilities need to be linked by well-designed trails for minibuses, canoe trucks, and other modes of transportation. To conserve the natural setting and to increase utility, safety, and control of such sites, planners should concentrate the facilities rather than disperse them.

Compatible neighbors include other organization camps, forests, and other extensive land uses. Second-home subdivisions, public parks and beaches, and ser-

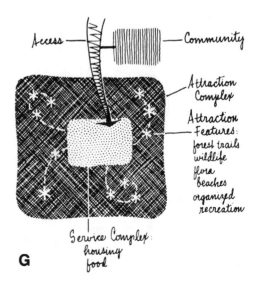

Figure 7-9. *Model G. For organization camping, functional relationships are similar to those in Model F but are even more tightly tied to the natural resource setting.*

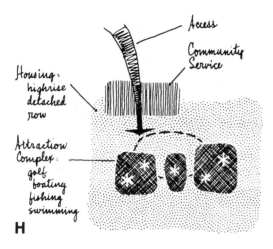

Figure 7-10. *Model H. For vacation home use, the natural resource setting, frequently water, is intimately linked to the physical development of housing.*

vice clusters are generally poor neighbors for organization camps.

Vacation Home Sites

Figures 7-10 through 7-13 (Model H) illustrate the basic functional relationships appropriate for vacation home use. In such destinations, the housing develop-

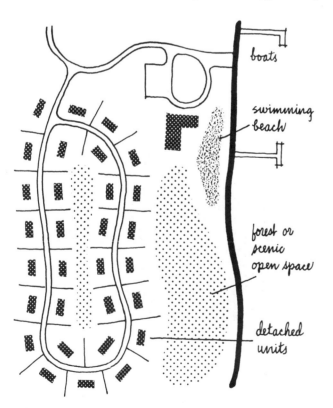

Figure 7-11. *Low-density vacation home layout with open space available along the waterfront.*

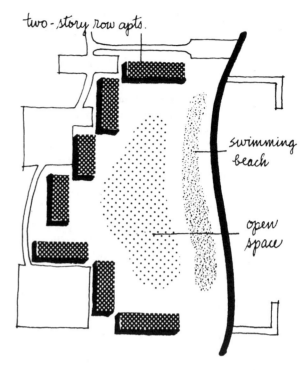

Figure 7-12. *Medium-density vacation homes, grouped in row units and maintaining an open waterfront commons.*

ment is frequently imbedded in the resource attraction or, often, surrounds it.

Settings and locations vary, but most vacationers seek proximity to water. Some prefer mountains; others wilderness; a few urban areas. For most of these users, scenic views, access to water, seclusion, woodland settings, and interesting topography are more important than cost.

Some market segments accept designs that concentrate housing in rows or "fourplexes" rather than detached units. Often management firms provide such amenities as swimming areas, golf courses, and exercise rooms. Another market segment uses mobile homes stationed at vacation home sites with similar amenities.

As with all water-oriented uses, designers must respect the limited waterfront. Illustrated in Figures 7-11 through 7-13 are three density patterns—all designed to leave the waterfront open. Generally, cars can be kept away from shore areas. Beach views and easy foot access are prime design considerations.

The relationship to community may be very weak or strong, depending on the activity patterns desired. Compatible nearby developments are shopping centers, cultural centers, and sports complexes. Heavy industry, camps, and intensive-use public parks are not the best neighbors.

Festival and Event Sites

Because they involve only temporary uses of sites, festivals and events are one of the most difficult classes of attractions for designers and developers to create and

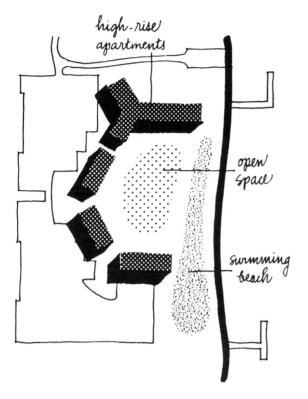

Figure 7-13. *High-density vacation homes achieved through high-rise units, again retaining an open space waterfront.*

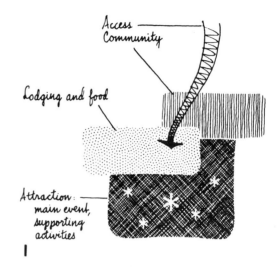

Figure 7-14. *Model I. For festival and event land use, the community is the dominant area. Large volumes of brief-stay visitors require special planning and management. Often, local tours are also involved.*

manage. Figure 7-14 (Model I) emphasizes that the event site, lodging and food, and support community are closely related to optimal functioning of such events.

Although their supporting services and attractions may be permanent, events thrust thousands of people into existing streets, parks, and other facilities for only short time periods. This process places great stress on the capacity and management control of event sites. Considerations of attractiveness, convenience, comfort, safety, toilet facilities, crowd control, and cleanliness are critical. As these attractions grow in popularity, catalytic action will be demanded of local leaders to pull together all agencies and organizations involved for planning and management. Although transitory, events require policies and decisions to satisfy the needs of thousands of visitors for short periods of time and not place undue stress on local communities.

Several longer-stay classes of tourist places also require community locations. Figure 7-15 (Model J) illustrates the functional concept for six classes of urban destination development. In all of the examples, community, lodging, and other services are closely tied to the attracting features.

Business Meeting Sites

Business meetings, conventions, conferences, seminars, workshops, marketing marts, and field trips constitute a major travel purpose. Previously considered func-

tions taking place only in larger cities, they are now increasingly important to medium-size and small cities. Whereas large conferences are more heavily promoted, those with fewer than two hundred attending are more numerous.

A shallow view of these events suggests only business and professional service needs. Certainly, well-designed meeting rooms with modern technical equipment are needed. But business and personal travel objectives are being combined, and more spouses and families are traveling with businesspersons. Itineraries include attractions in the surrounding areas as well as at the meeting places.

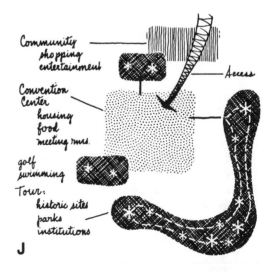

Figure 7-15. *Model J. This functional model represents relationships for several kinds of attractions: business and meeting places, gaming centers, sports arenas, trade centers, science and technology centers, and theme parks.*

Too often, interesting food services, shops, entertainment, and museums are not designed for pedestrian access near convention centers. Shuttle linkage to surrounding attractions and tours to interesting places are also design, planning, and development needs.

Gaming Centers

Gambling is emerging as a major expansion of tourism development. Although the major focus at these centers may be on gambling, other activities are also in demand. For example, evening entertainment and surrounding natural resource attractions are as important for Las Vegas as casinos. Millions of visitors tour Hoover Dam and the surrounding attractions of Lake Mead, the Grand Canyon, local ski areas, and Native American reservations. Again, the design and integration of complexes is as important as the design of individual attraction features.

Sports Arenas

Surrounded by huge parking areas, sports arenas are great land consumers. Nearness to community amenities and access is important. In recent years, however, arenas have often usurped prime riverfront or historic areas that could otherwise have been attractions themselves. When a development is made solely because of its general location and price, it sometimes destroys other potential amenities. Certainly, an integrated location and development approach is needed.

Trade Centers

Along with convention centers, trade marts are increasingly important travel targets that seldom cater to single-purpose markets. Nearby amenities are worthy of design and development consideration.

Science and Technology Centers

As scientific and technological activity becomes more complex, the need to travel increases, in spite of expanding telecommunications. Places such as the NASA/Johnson Space Center, near Houston, have become focal points for much business travel. Often spouses and families accompany businesspersons on such trips and seek diversions in the surrounding areas. Exhibits and interpretation become important attractions for them. The land and building design for this mix of functions requires special study for each site.

Theme Parks

Perhaps no other attraction class has increased pleasure travel more than theme parks. They represent a highly specialized aspect of land use and planning. The

feeding, circulation, and entertainment of great numbers of people demand highly skilled technicians and creative designers working together. Safety, sanitation, and crowd control are paramount.

Because the travel markets for theme parks usually come from a radius of less than 300 miles, repeat visits demand change. Some operators believe they must add a new ride or attraction every two years. Several theme parks that were initially developed for children have added entertainment, activities, and food services for older populations. Because of their innovative design and responsible management, theme parks have established themselves as high-quality travel attractions. Despite criticism by some environmentalists, there is no evidence that theme parks are not just as important to society as national parks.

SPECIAL CASE: NATIONAL PARKS

Because the design and management of national parks have been so controversial, they deserve special comment. For many years the national park agencies of several countries have wrestled with the dual policies of resource protection and public use. Books, special studies, and heated discussions and writings have perpetuated the myth that these are incompatible objectives. The thesis here is that the problems usually cited are the result of poor design and poor management, both of which can be changed.

Model K, shown in Figure 7-16, illustrates key functional relationships for national park areas. Following is

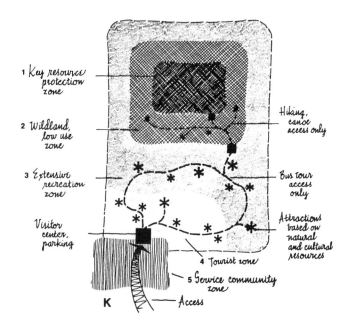

Figure 7-16. *Model K. A concept for planning a national park into five zones. Most visitor services (lodging, food, travel services, entertainment, shopping) should be located some distance from prime resources, preferably in a community at the edge of the park.*

interesting food services, shops, entertain-
museums are not designed for pedestrian
convention centers. Shuttle linkage to sur-
tractions and tours to interesting places are
planning, and development needs.

enters

g is emerging as a major expansion of tourism
t. Although the major focus at these centers
gambling, other activities are also in demand.
e, evening entertainment and surrounding nat-
e attractions are as important for Las Vegas as
ions of visitors tour Hoover Dam and the sur-
tractions of Lake Mead, the Grand Canyon,
as, and Native American reservations. Again,
nd integration of complexes is as important as
f individual attraction features.

enas

led by huge parking areas, sports arenas are
consumers. Nearness to community amenities
is important. In recent years, however, are-
ten usurped prime riverfront or historic areas
otherwise have been attractions themselves.
velopment is made solely because of its gen-
on and price, it sometimes destroys other
enities. Certainly, an integrated location and
t approach is needed.

ters

ith convention centers, trade marts are in-
mportant travel targets that seldom cater to
ose markets. Nearby amenities are worthy of
development consideration.

d Technology Centers

ntific and technological activity becomes
lex, the need to travel increases, in spite of
elecommunications. Places such as the NASA/
ace Center, near Houston, have become focal
much business travel. Often spouses and fam-
pany businesspersons on such trips and seek
n the surrounding areas. Exhibits and inter-
ecome important attractions for them. The
uilding design for this mix of functions
cial study for each site.

rks

no other attraction class has increased plea-
more than theme parks. They represent a
alized aspect of land use and planning. The

feeding, circulation, and entertainment of great numbers
of people demand highly skilled technicians and creative
designers working together. Safety, sanitation, and crowd
control are paramount.

Because the travel markets for theme parks usually
come from a radius of less than 300 miles, repeat visits
demand change. Some operators believe they must add
a new ride or attraction every two years. Several theme
parks that were initially developed for children have
added entertainment, activities, and food services for
older populations. Because of their innovative design
and responsible management, theme parks have established themselves as high-quality travel attractions.
Despite criticism by some environmentalists, there is no
evidence that theme parks are not just as important to
society as national parks.

SPECIAL CASE: NATIONAL PARKS

Because the design and management of national parks
have been so controversial, they deserve special comment.
For many years the national park agencies of several countries have wrestled with the dual policies of resource protection and public use. Books, special studies, and heated
discussions and writings have perpetuated the myth that
these are incompatible objectives. The thesis here is that
the problems usually cited are the result of poor design
and poor management, both of which can be changed.

Model K, shown in Figure 7-16, illustrates key functional relationships for national park areas. Following is

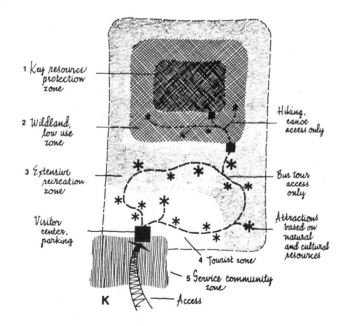

Figure 7-16. *Model K. A concept for planning a national
park into five zones. Most visitor services (lodging, food,
travel services, entertainment, shopping) should be located
some distance from prime resources, preferably in a community at the edge of the park.*

sions, and extensive-use public parks. Certain natural
resources, such as a warm climate, may encourage the
creation of new resort cities. In such cases, designers and
planners must encourage the protection of basic attraction resources.

In the United States and Canada, ranch resorts with
cowboy and western themes continue to gain in popularity.
These vary along a scale from operating farms and ranches
to contrived complexes in ranch-like settings. Farm and
ranch businesses maintain their primary functions, perhaps
with only slight remodeling of the owners' homes to
accommodate visitors, allowing merely a few guests at
selected seasons. At the other end of the scale are huge
complexes with lodging, food service, country and western
entertainment, and other recreational amenities, such as
golf courses, tennis courts, pools, and equestrian trails.
Such complexes require new design input.

Camping Areas

Figure 7-8 (Model F) illustrates two classes of development—natural resources camping areas and hunting and
water sports areas.

Adults and families continue to use RVs to get to these
destinations. Older and retired RV users who spend the
winter in warm climates usually prefer community amenities nearby. Park sites often contain central activity
rooms, golf courses, hiking trails, and bicycle trails.

Frequently, package tours of these sites expand opportunities for enjoyable and healthful activities. In northern
climates, summer use of natural resource attractions continues to be popular for all age groups. Those who design
such places must try to provide attractive settings for specialized market segments that stay at a destination for an
extended period of time.

Hunting and Water Sports Areas

Those who engage in hunting and fishing expeditions
may camp near a resource area or occupy commercial
lodging in the nearest community. All facilities close to
or within the resource area should be designed to be
compatible with the special resource features and to
emphasize environmental sustainability.

Organization Camping Sites

For many years young people and adults have engaged
in forms of camping exclusive to each group. The functional relationships relevant to this purpose are shown in
Figure 7-9 (Model G). Examples include camps led by the
Boy Scouts, the Girl Scouts, Young Men's Christian Associations, Young Women's Christian Associations, and 4-H
clubs, as well as by church and conservation groups. In
addition to the main goal of living close to nature, such
groups may have educational, religious, or fraternal objec-

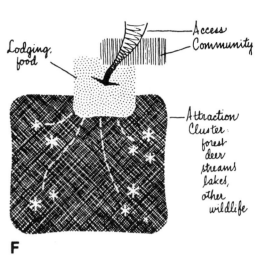

Figure 7-8. *Model F. Functional relationships are quite
similar for longer-stay camping areas and hunting and
water sports areas. Natural resource settings dominate the
attractions.*

tives. The development of organization camps is promoted
heavily by exponents of outdoor and conservation education, now known as "environmental education."

Although it is a key destination activity, organization
camping land use is conceptually the reverse of many
others. Attraction clusters surrounding a core of facilities
need to be linked by well-designed trails for minibuses,
canoe trucks, and other modes of transportation. To
conserve the natural setting and to increase utility, safety,
and control of such sites, planners should concentrate
the facilities rather than disperse them.

Compatible neighbors include other organization
camps, forests, and other extensive land uses. Second-
home subdivisions, public parks and beaches, and ser-

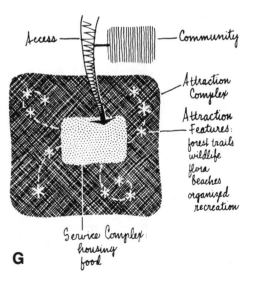

Figure 7-9. *Model G. For organization camping, functional
relationships are similar to those in Model F but are even
more tightly tied to the natural resource setting.*

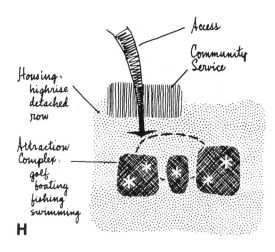

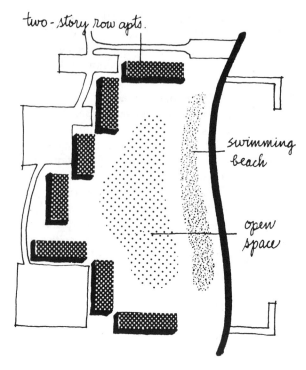

Figure 7-10. *Model H. For vacation home use, the natural resource setting, frequently water, is intimately linked to the physical development of housing.*

vice clusters are generally poor neighbors for organization camps.

Vacation Home Sites

Figures 7-10 through 7-13 (Model H) illustrate the basic functional relationships appropriate for vacation home use. In such destinations, the housing develop-

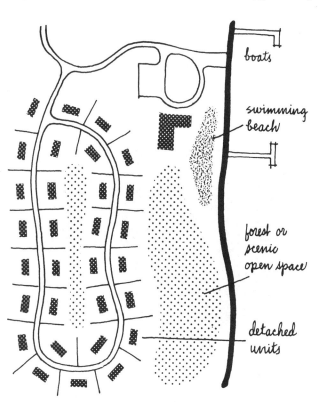

Figure 7-11. *Low-density vacation home layout with open space available along the waterfront.*

Figure 7-12. *Medium-density vacation homes, grouped in row units and maintaining an open waterfront commons.*

ment is frequently imbedded in the resource attraction or, often, surrounds it.

Settings and locations vary, but most vacationers seek proximity to water. Some prefer mountains; others wilderness; a few urban areas. For most of these users, scenic views, access to water, seclusion, woodland settings, and interesting topography are more important than cost.

Some market segments accept designs that concentrate housing in rows or "fourplexes" rather than detached units. Often management firms provide such amenities as swimming areas, golf courses, and exercise rooms. Another market segment uses mobile homes stationed at vacation home sites with similar amenities.

As with all water-oriented uses, designers must respect the limited waterfront. Illustrated in Figures 7-11 through 7-13 are three density patterns—all designed to leave the waterfront open. Generally, cars can be kept away from shore areas. Beach views and easy foot access are prime design considerations.

The relationship to community may be very weak or strong, depending on the activity patterns desired. Compatible nearby developments are shopping centers, cultural centers, and sports complexes. Heavy industry, camps, and intensive-use public parks are not the best neighbors.

Festival and Event Sites

Because they involve only temporary uses of sites, festivals and events are one of the most difficult classes of attractions for designers and developers to create and

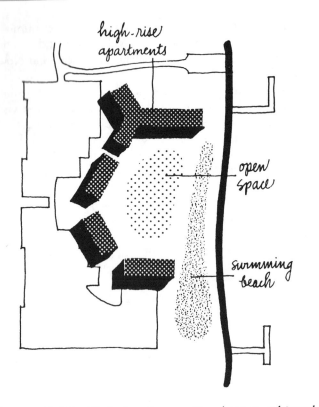

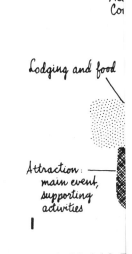

Figure 7-13. *High-density vacation homes achieved through high-rise units, again retaining an open space waterfront.*

Figure 7-14. *Model I. F... community is the domin... stay visitors require spe... Often, local tours are also...*

manage. Figure 7-14 (Model I) emphasizes that the event site, lodging and food, and support community are closely related to optimal functioning of such events.

Although their supporting services and attractions may be permanent, events thrust thousands of people into existing streets, parks, and other facilities for only short time periods. This process places great stress on the capacity and management control of event sites. Considerations of attractiveness, convenience, comfort, safety, toilet facilities, crowd control, and cleanliness are critical. As these attractions grow in popularity, catalytic action will be demanded of local leaders to pull together all agencies and organizations involved for planning and management. Although transitory, events require policies and decisions to satisfy the needs of thousands of visitors for short periods of time and not place undue stress on local communities.

Several longer-stay classes of tourist places also require community locations. Figure 7-15 (Model J) illustrates the functional concept for six classes of urban destination development. In all of the examples, community, lodging, and other services are closely tied to the attracting features.

Business Meeting Sites

Business meetings, conventions, conferences, seminars, workshops, marketing marts, and field trips constitute a major travel purpose. Previously considered func-

Too often... ment, an... access nea... rounding... also desig...

Gaming...

Gamblin... developme... may be on... For examp... ural resou... casinos. M... rounding a... local ski a... the design... the design...

Sports A...

Surrou... great lan... and acces... nas have o... that coul... When a d... eral locat... potential a... developme...

Trade Ce...

Along... creasingly... single-pur... design an...

Science a...

As sci... more com... expanding... Johnson S... points for... ilies acco... diversions... pretation... land an... requires s...

Figure 7-15. *Model J. relationships for several meeting places, gaming ters, science and techno...*

Theme P...

Perhap... sure trave... highly p...

a descriptive scenario that could be the foundation for both greater public use of national parks and greater protection of basic resources. Its fundamental principle of zoning was expressed several years ago in Canada as national park policy that recommended a zoning plan with a wilderness area linked to a permanent townsite (Canada 1969). A similar concept was set forth in a document of the International Union for Conservation of Nature and Natural Resources: "If parks can be classified into different zones managed to meet different sets of objectives, the tension between perpetuation and use will be minimized" (Forster 1973, 49). Landscape architect Richard R. Forster's concept includes three zones: a protected resource core limited to scientists only, a partial reserve where visitors would be permitted but restricted to authorized trails and parking areas, and an outer zone for recreation and access by tourists. B. K. Downie (1984) has suggested a five-zone concept, with zones ranging from special preservation areas to visitor services.

The following scenario builds on the work of these authors and suggests a five-zone pattern as shown in the model. It begins with certain basic assumptions:

1. National park boundaries are political, not ecological. The final boundary established is a legislated line that circumscribes an area of varied resources, only some of which are ecologically or culturally significant, rare, or fragile.
2. Within these boundaries the land characteristics are not homogeneous. Some portions may be rare, fragile, or unusual; other portions may have no unusual characteristics and may already show development by man.
3. Policy decrees public use of the land because society can benefit from contact with and understanding of the protected resources. Visitors are enriched by exposure to natural and cultural phenomena.
4. An analysis of visitors has shown that, whereas many can share the experience of key elements of the parks, the market tends to segment itself in its use of supportive services (food, lodging, retail sales, entertainment).
5. Some parks attract many more millions of visitors than others because of proximity to population centers, ease of access from markets, or popularity of features.

Certain planning and design concepts can be based on these assumptions. Through land resource study and analysis of a park and its surrounding areas, several zones can be identified. These zones provide designers, developers, and managers with strong foundations for policy. Instead of a blanket policy that erroneously assumes that everything within park boundaries requires the same rules, each of the following zones can have its own set, based on ecological and biological foundations as well as on market considerations:

1. *Key resource protection zone.* Qualified biologists, landscape architects, historians, and archaeologists can identify the prime resources of the park, including scenic resources; rare plants; important animals and habitats; and areas of architectural, historic, or archaeological value that cannot withstand visitor intervention. Generally, these are the features that stimulated the creation of the park in the first place. This zone should be reserved for scientific use.
2. *Wildland/low use zone.* Within the park, other lands may be less valuable for protection but still contain resources that would be disturbed if roads and facilities were introduced. These areas should constitute a roadless zone with limited visitor access, much like areas presently governed by U.S. wilderness policy.
3. *Extensive recreation zone.* Specialists can identify specific areas outside the wildland zone that are extensive and stable enough to support visitor use. Extensive recreation, such as hunting, fishing, and camping, can be provided in accordance with proper planning and design principles. The extent of this development must be in balance with the resource base.
4. *Tourist zone.* Special sites and travelways for motorcoaches, monorails, or other forms of mass transportation can be installed for visitor use. Proper design of facilities to handle mass use can both protect resources and allow public enrichment. For example, travelways can penetrate restricted zones, permitting visitors to see, but not come in contact with, special resources. Such travelways should not be open to personal vehicles. Special turnouts, overlooks, and interpretation centers along the way can contain educational exhibits, literature, slide presentations, and lecture rooms.
5. *Service community zone.* At the edge of the park, either just inside or just outside its boundaries, a service community zone can operate independently as a city. All development can flow according to market demand from luxury to economy. Today, enough is known about urban design to plan and create attractive, functional, and market-oriented communities. Although a new town may be necessary, it is far better to expand one that has an established infrastructure and management policy.

Although it certainly does not address all national park design and management issues, the five-zone planning concept can resolve many management problems. Its implementation would allow millions of visitors to use a park without damaging its resources. The merit of this scenario is demonstrated in several areas where at least some of these principles are practiced.

For example, numbers of red-cockaded woodpeckers, Florida sandhill cranes, round-tailed water rats, and American alligators have actually increased in number at the Okefenokee National Wildlife Refuge at the same time the number of visitors has also increased (Gunn 1979). This result is because of the prohibition of personal cars and the use of a rubber-tired interpretive visitor train, which allows visitor enrichment while reducing vandalism in the area and disturbance to wildlife. The concept contributes to what Joseph L. Sax (1980, 111) has identified as the four key meanings of U.S. national parks: places where recreation reflects the aspirations of a free and independent people; object lessons for a world of limited resources; great laboratories of successful natural communities; and living memorials of human history on the American continent.

When a national, state, or provincial park is zoned in this manner, visitors need to be informed about the zone characteristics and boundaries. Upon entering the park, visitors should be given maps describing the resources, policies, and functions of each zone. Color coding on the map should correspond to well-designed markers on trails and roads at zone junctures.

Whether this or some other scheme is adopted for national park planning and design, one principle should be clear, as stated by Sax (1980, 103): "Most conflict over national park policy does not really turn on whether we ought to have nature reserves (for that is widely agreed), but on the uses that people will make of those places, which is neither a subject of general agreement nor capable of resolution by reference to ecological principles." Design and management should put to rest the dichotomy of resource protection versus public use and strive for the effective operation of both. When designers, planners, managers, the commercial tourism sector, and residents begin to apply their talents generously, forcefully, and cooperatively to this end, they can demonstrate that protection and public use can be a happy marriage rather than a cause for divorce.

Stronger understanding and cooperation between national parks and the private tourism sectors in the United States is being fostered by a new management tool, the Money Generation Model (MGM). This is an economic impact model that demonstrates clearly how these two development sectors are intertwined.

The MGM uses park visitation data as a basis for estimating economic impacts for three revenue sources: park-related tourist spending, money spent by the park itself, and money spent by nonlocal parties that would not have been spent were it not for the park (Hornback 1991). When applied to Minute Man National Historical Park in Massachusetts, the MGM model produced several positive results beyond statistics on economics (Gall 1991): local acceptance of the General Management Plan (GMP) and legislative proposals for land acquisition; strengthening relationships with local historical and cultural institutions; and building relationships with local business community. These benefits helped to eliminate local conflict, improve communication, and build common agendas and support networks. This example of a management tool demon-strates how one management approach, here an economic model, has far greater value than economic measures alone in stimulating greater cooperation between the tourist business sector and the public sector.

SPECIAL CASE: ECOTOURISM

The modern thrust of ecotourism development is very similar to trends in national park spatial development that stressed separation of services from protected areas. On the positive side, this form of travel reflects new interest in nature and culture among populations worldwide. But ecotourism brings with it special requirements for planning and development of the supply side—attractions, transportation and access, services, information and promotion. Two major issues must be addressed.

First, the question arises of how resources can be protected at the same time visitor numbers are increasing. A first step toward the solution of this question is to obtain an objective scientific analysis of a site's key resource characteristics, especially their rarity, fragility and tolerance for human contact. After the ecosystem land area for plant and animal life has been determined, specialists can recommend the limits for visitor encroachment. Too close contact with wild animals can often create a dependency and bonding because of which visitors tend to treat the animals as pets, forgetting that they are still wild and can quickly revert to behavior that threatens the lives of humans. Many years ago, such a people–wildlife dependence developed in Yellowstone National Park, with brown and grizzly bears feeding on hotel garbage and handouts from tourists. Because of consequent attacks on visitors, policy now requires separation between visitors and bears. As ecotourism becomes more popular, the design of trails, overlooks, and interpretive centers will require special study and execution for each site and area so that its plant and animal species can be perpetuated at the same time as more visitors have the opportunity to be enriched by these resources.

A second major issue centers on facilities. To house, feed, and provide programs for visitors, extreme care in location and site design must be taken. Some facilities that claim to be ecolodges are actually destructive of native plants, animals, and cultural settings. As described in Chapter 8, there is an ethical issue involved in placing buildings, walks, drives, and volumes of visitors directly on sites that cannot support these burdens without serious damage to the aesthetics of their views and the habitats of their plant and animal life.

A review of diagram Figure 7-17 (Model L) can assist developers of ecotourism in their placement of facilities and services. The five options for locating prime facilities for lodging and food service are as follows:

A *Urban.* For many travelers, a tour into a resource area on a day-use basis is acceptable. The nearby major city provides the support facilities in a range of market preference, from luxury to economy. The community setting is suitable because it is easily accessible and contains other desired amenities. The ecotour may be self-guided, offered by the management of the resource area, or offered by a private firm. The advantage of this option is that it does not require the resource area manager to provide tourist services except suitable trails and interpretation within the area.

B *Village.* This option places the lodging and other support facilities in a small town closer to the resource area. The town may not have the amenities of a larger city and thus may require shuttle linkage with a larger city. The development of ecolodges here requires acceptance from community leaders to make sure they are compatible. Again, visitors may uses ecotours offered by the resource area management or by private firms.

ECOLODGE SPATIAL PATTERNS

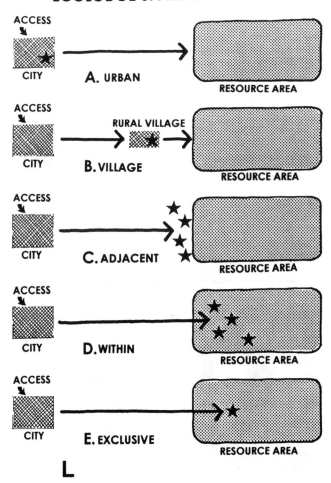

Figure 7-17. *Model L. Ecolodge spatial patterns. Several options for lodging and travel service locations. Least threatening to natural and cultural resources are concepts A, B, and C.*

C *Adjacent.* A pattern frequently used surrounding national parks in Africa is the development of lodges and tourist facilities directly adjacent to the parks. Safari lodges located near Kruger National Park in South Africa provide basic services, their own ecotours, and linkage with tours inside the park. As in the options above, this pattern relieves the park management from providing these services and allows them to retain control of visitor impact on the park.

D *Within.* Hindsight into situations at national parks in the United States and Canada suggests that in the past, when visitor volumes were small, within-park facilities were a suitable arrangement. However, as volumes reach the millions, the magnitude of the transportation and urban-like infrastructure (water, waste, police, fire control) required for such facilities creates a conflict with the purpose of the park. If their location is at the edge of a park, where such damage can be minimized, then adaptation to tourism may be suitable. One advantage to facilities within is the revenue paid to the park by private concessioners.

E *Exclusive.* In some cases, owner-developers create ecolodges on their own tracts of land that are large enough to provide the plant and animal resources desired by ecotourists. An example is the Kari Resort, located in a forest area south of Perth, Australia, where a small lake was formed by damming a stream. For such an enterprise to succeed, the developer must carry the burden of resource as well as resort management, including all provisions for infrastructure.

Every local case must be evaluated as to its specific conditions, but these patterns provide the basis for making development decisions.

SPECIAL CASE: COASTS, WATERFRONTS

Probably no other landform is as compelling for tourism as the waterfront, especially the coastal zone. This zone is neither land nor water but a special amalgam with more powerful properties than the sum of the two. Places where the land and sea meet not only hold contemporary recreational interest but also evoke nostalgia. "We still like to go beach-combing, returning to primitive act and mood. When all the lands will be filled with people and machines, perhaps the last need and observance of man will be, as it was in the beginning, to come down and experience the sea" (Sauer 1967, 310).

But the linear coast presents special tourism design challenges. For descriptive purposes, it can be divided into four ribbons, each accorded a special use and, thus, special design needs (Figure 7-18) (Gunn 1972):

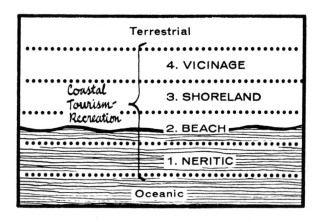

Figure 7-18. *Linear coastal zones. The four zones parallel to the beach have somewhat different potential for adaptation to tourism. The most stable area for development lies well back from the water's edge.*

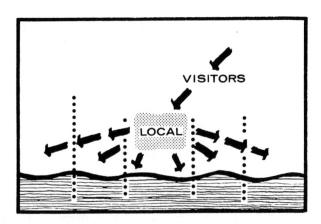

Figure 7-21. *Tourist access through a city. Generally favored locations for coastal tourism development are in or near a service community. Such development must be planned in coordination with other waterfront uses.*

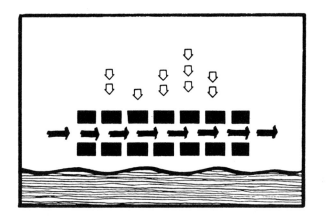

Figure 7-19. *Traditional coastal development. In the past, highways too close to the shoreline have restricted proper use of this valuable and limited asset. This pattern blocks vistas and access from the hinterland.*

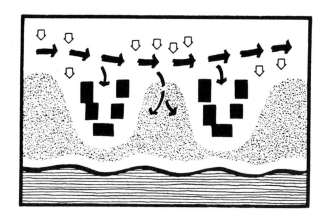

Figure 7-22. *Building envelopes. For many years, planners have advocated the land use pattern of building envelopes along coasts. This pattern protects open waterfront and yet allows high-density development.*

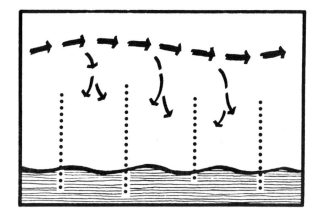

Figure 7-20. *Desirable coastal plan. By placement of the road behind most development, better land use is assured and conflict between pedestrians and automobiles is avoided.*

Figure 7-23. *Industrial waterfront tourism. Along harbors, many types of water-oriented industries, such as shipping and commercial fishing, can also double as tourist attractions if properly planned and managed.*

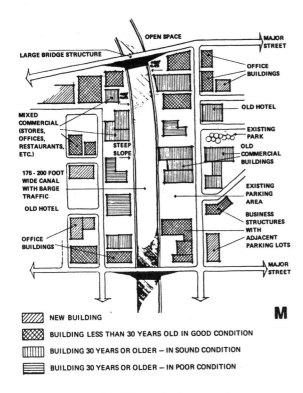

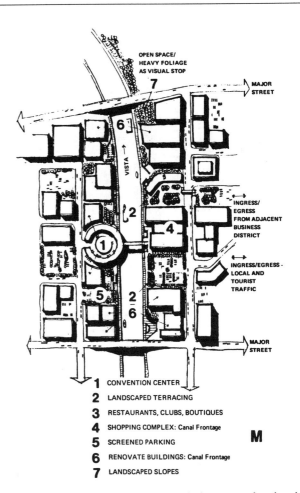

Figure 7-24. *Model M. Typical urban canal setting.*

1 CONVENTION CENTER
2 LANDSCAPED TERRACING
3 RESTAURANTS, CLUBS, BOUTIQUES
4 SHOPPING COMPLEX: Canal Frontage
5 SCREENED PARKING
6 RENOVATE BUILDINGS: Canal Frontage
7 LANDSCAPED SLOPES

M

Figure 7-26. *Model M. Recommended concept for development solution.*

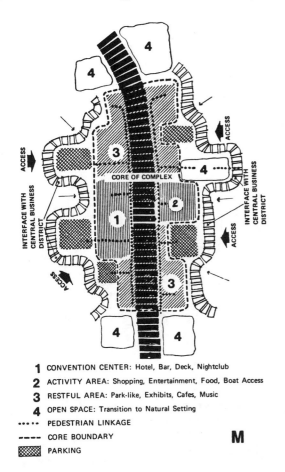

1 CONVENTION CENTER: Hotel, Bar, Deck, Nightclub
2 ACTIVITY AREA: Shopping, Entertainment, Food, Boat Access
3 RESTFUL AREA: Park-like, Exhibits, Cafes, Music
4 OPEN SPACE: Transition to Natural Setting
···· PEDESTRIAN LINKAGE
---- CORE BOUNDARY
▨ PARKING

M

Figure 7-25. *Model M. Proposed functional land use diagram.*

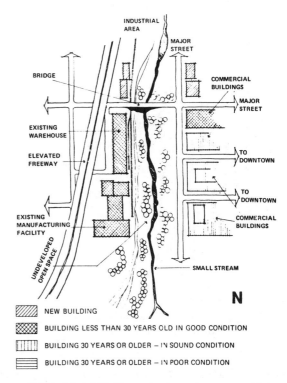

▨ NEW BUILDING
▨ BUILDING LESS THAN 30 YEARS OLD IN GOOD CONDITION
▥ BUILDING 30 YEARS OR OLDER — IN SOUND CONDITION
▤ BUILDING 30 YEARS OR OLDER — IN POOR CONDITION

N

Figure 7-27. *Model N. Typical urban stream corridor.*

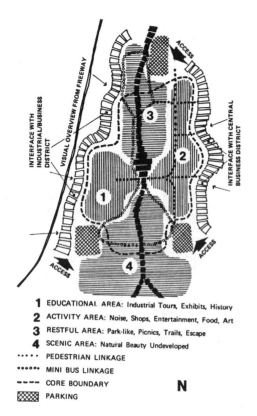

Figure 7-28. *Model N. Proposed functional land use diagram.*

1 EDUCATIONAL AREA: Industrial Tours, Exhibits, History
2 ACTIVITY AREA: Noise, Shops, Entertainment, Food, Art
3 RESTFUL AREA: Park-like, Picnics, Trails, Escape
4 SCENIC AREA: Natural Beauty Undeveloped
····· PEDESTRIAN LINKAGE
•••••• MINI BUS LINKAGE
- - - CORE BOUNDARY
▨▨▨ PARKING   N

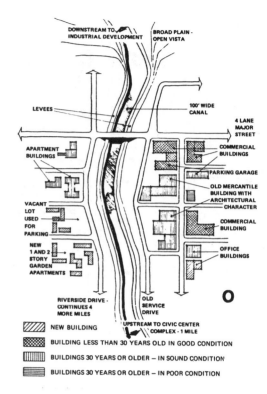

Figure 7-30. *Model O. Typical urban river setting.*

▨▨▨ NEW BUILDING
▩▩▩ BUILDING LESS THAN 30 YEARS OLD IN GOOD CONDITION
▥▥▥ BUILDINGS 30 YEARS OR OLDER — IN SOUND CONDITION
▤▤▤ BUILDINGS 30 YEARS OR OLDER — IN POOR CONDITION

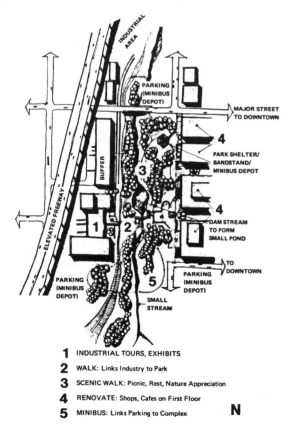

1 INDUSTRIAL TOURS, EXHIBITS
2 WALK: Links Industry to Park
3 SCENIC WALK: Picnic, Rest, Nature Appreciation
4 RENOVATE: Shops, Cafes on First Floor
5 MINIBUS: Links Parking to Complex   N

Figure 7-29. *Model N. Recommended concept for development solution.*

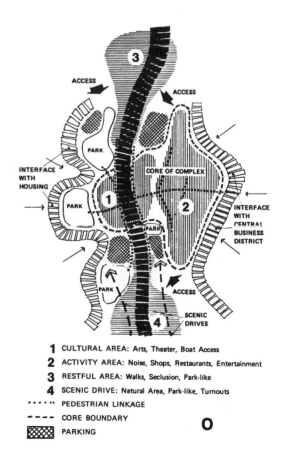

1 CULTURAL AREA: Arts, Theater, Boat Access
2 ACTIVITY AREA: Noise, Shops, Restaurants, Entertainment
3 RESTFUL AREA: Walks, Seclusion, Park-like
4 SCENIC DRIVE: Natural Area, Park-like, Turnouts
····· PEDESTRIAN LINKAGE
- - - CORE BOUNDARY
▨▨▨ PARKING   O

Figure 7-31. *Model O. Proposed functional land use diagram.*

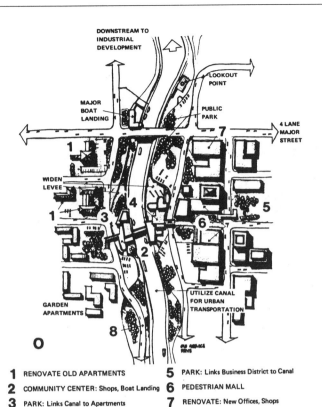

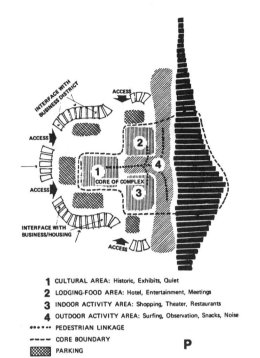

1 CULTURAL AREA: Historic, Exhibits, Quiet

2 LODGING-FOOD AREA: Hotel, Entertainment, Meetings

3 INDOOR ACTIVITY AREA: Shopping, Theater, Restaurants

4 OUTDOOR ACTIVITY AREA: Surfing, Observation, Snacks, Noise

•••• PEDESTRIAN LINKAGE

---- CORE BOUNDARY

▨ PARKING

Figure 7-34. *Model P. Proposed functional land use diagram.*

1 RENOVATE OLD APARTMENTS

2 COMMUNITY CENTER: Shops, Boat Landing

3 PARK: Links Canal to Apartments

4 COMMERCIAL RIVERFRONT COMPLEX

5 PARK: Links Business District to Canal

6 PEDESTRIAN MALL

7 RENOVATE: New Offices, Shops

8 SCENIC DRIVE ON LEVEE

Figure 7-32. *Model O. Recommended concept for development solution.*

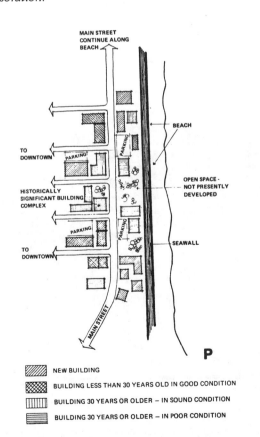

NEW BUILDING

BUILDING LESS THAN 30 YEARS OLD IN GOOD CONDITION

BUILDING 30 YEARS OR OLDER – IN SOUND CONDITION

BUILDING 30 YEARS OR OLDER – IN POOR CONDITION

Figure 7-33. *Model P. Typical urban waterfront.*

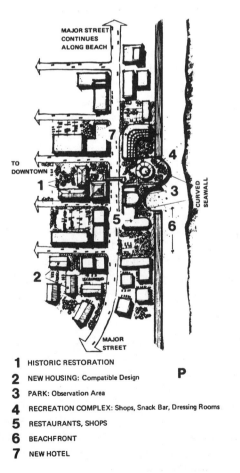

1 HISTORIC RESTORATION

2 NEW HOUSING: Compatible Design

3 PARK: Observation Area

4 RECREATION COMPLEX: Shops, Snack Bar, Dressing Rooms

5 RESTAURANTS, SHOPS

6 BEACHFRONT

7 NEW HOTEL

Figure 7-35. *Model P. Recommended concept for development solution.*

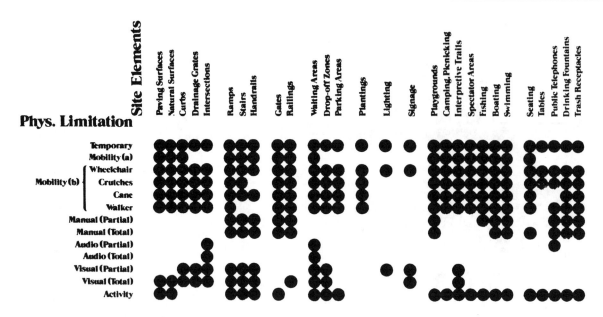

Figure 7-36. *Handicap–site relationship. Black dots represent areas where site elements may cause physical barriers for handicapped people, requiring special design solutions.*

1. *Neritic.* This ecological "near-shore" zone spreads from the continental shelf to the beach. It is the richest zone for fishing and often contains interesting sandbars and reefs. It is well suited to cruising, sailing, and travel to nearby islands. Visual contact is predominantly with the sea.
2. *Beach.* The beach zone reaches into the water and onto the land. If wide and sandy, it supports the most popular forms of water-based recreation, as well as relaxing, sand-castle-building, beach sports, people-watching, and action and nature photography.
3. *Shoreland.* Behind the beach lies the setting for camping, picnicking, and hiking. The shoreland zone may also support hotels and other service businesses. Visual linkage between land and sea is important.
4. *Vicinage.* The marine coastal backland is the setting for tourist businesses and vacation homes. In this vicinage zone, the coastal scenery is often enhanced by variations in topography and vegetative cover. Nearness and access to the sea are more important than visual linkage.

Historically, the first tourism development along coasts and inland waterfronts has usually been of roads parallel and close to the beach (Figure 7-19). Such roads fulfilled the need for access and resulted in building between the road and the beach. In popular beach areas, massive construction walled off access, views, and air circulation from the backlands. Resorts and vacation homes situated behind the roads faced concrete backsides and garbage containers.

A wiser plan would have placed main service roads farther away from the beach, providing access to all four zones (Figure 7-20). Such a plan would have addressed the fact that primary access comes not along the beach but from the inland region, perpendicular to the beach. Such a design would also have avoided circulation conflict between pedestrians going from the homes to the beach and automobiles traveling up and down the shoreline. The coastal zone, despite its uniform physical characteristics parallel to the water's edge, is not uniform in demand because of its coastal communities and primary access routes to them (Figure 7-21).

Figure 7-37. *General site accessibility. Design for the handicapped requires integration of the complete complex, including bus stop shelters and access to buildings.*

For many years designers and planners have proposed the construction of clusters of buildings or high-rise envelopes along coastal zones (Figure 7-22). Had this been the pattern at Miami Beach and Waikiki Beach, much of the present blockage of waterfront views and access at those beaches could have been avoided. Because implementation of this proposal would lead to greater and better use of waterfront and backland, it is probably a more economic method for future developments.

In the redevelopment of waterfronts, the smells, sounds, and views of remaining port operations are often interesting to visitors. Derelict properties in between these operations can be razed and developed into interesting park spaces (Figure 7-23).

Many cities seek ways of rejuvenating their downtown areas for visitors and residents. If they have water resources, there is a much greater potential for successful redevelopment. Research has shown that development of water resources downtown is influenced greatly by the stability of the water level. This finding is especially true of urban rivers, as is demonstrated by the successful River Walk in the core of San Antonio, Texas (Gunn, Reed, and Couch, 1972). Following an investigation of the potential of many urban waterfronts, Gunn, Hanna, Perenzin, and Blumberg (1974) recommended a three-phase process for their redevelopment: investigate motivating factors (who

is supportive); analyze site factors (physical assets limitations); and evaluate important external factors (price, regulations, ownership, access).

As an aid to visualizing waterfront development opportunities, four hypothetical model sites are illustrated (Figures 7-24 through 7-35). The diagram given for each model represents a typical urban waterfront situation. These diagrams were studied by landscape architects Clare A. Gunn and Robert Green, who created a functional diagram and a concept plan for each (Gunn et al. 1974).

SPECIAL POPULATIONS

Greater recognition by society and more abundant legislation concerning the rights and interests of special populations have necessitated increasing design responsibility. Especially needed are designs that make travel easier for the mentally ill, mentally retarded, disabled, aged, and sensory-impaired, who represent growing travel markets. Much progress has been made in recent years. To continue this progress, barrier-free renovations and new designs of tourist-oriented businesses, attractions, and transportation that take into account the needs of special populations must be a part of all tourism development.

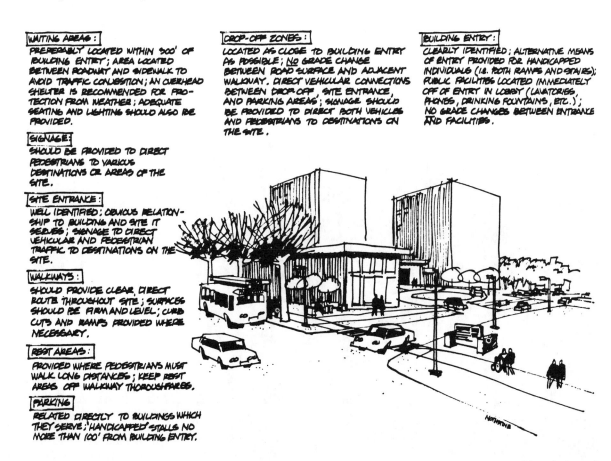

Figure 7-38. *Waiting areas. Bus and car stops must be designed to accommodate the special needs of handicapped persons.*

If designers and developers are to meet the needs of special travel populations, they should avail themselves of standards and guides for building and site design. For example, *Barrier Free Site Design* (American Society of Landscape Architects and U.S. Department of Housing and Urban Development 1977, 20), a publication that represents the state of the art for handicap design, provides a multitude of specifications, recommendations, and examples:

1. Special transportation facilities should be provided for people with restricted use of the exterior environment. Care should be taken to separate various modes of transportation, where practical, because their points of intersection are usually confusing, dangerous, and delaying. Vehicular traffic should be separated as much as possible from bicycle traffic, and both should be distanced from pedestrian traffic.
2. In general, access between transportation facilities and buildings should be smooth and free of barriers that may prove impossible for physically restricted people to negotiate. Paving surfaces should be hard and relatively smooth, curbs should have ramped cuts, walks should be sufficiently wide to accommodate two-way traffic, and entrance walks to buildings should slope gently to the platform before the doors. In buildings where exterior stairs are required, at least one major entrance should be served by a ramp as well.
3. Doors into public buildings should preferably be activated by automatic opening devices. If the cost is prohibitive, horizontal levers or through bars should be installed on the doors.
4. Public conveniences, such as restroom facilities, drinking fountains, telephones, elevators, and waiting areas should be well organized and located in close proximity to building entrances to allow people with physical limitations to gain access to necessary facilities with a minimal amount of hardship or embarrassment.

Some excerpts from *Barrier Free Site Design* are illustrated in Figures 7-36 through 7-38.

CONCLUSIONS

The significance of this chapter is that it provides land use patterns for the typical kinds of tourism destinations. The physical relationship between the community and other elements of destinations is an important factor to be considered by all who seek tourism growth.

Using the destination as the planning unit emphasizes the area around as well as within the community. Because such external lands contain major natural resources important to tourism, they must be part of community tourism decisions. Even though most tourism dollars are spent in service communities, these expenditures would not be made if it were not for the attractions surrounding as well as within the community. By focusing on the two classes of tourism—touring-circuit and longer-stay—planners and developers can use the several spatial patterns provided in their decision making. Communities and destinations gain most when touring circuits and longer-stay tourism are planned together.

Local designers and developers of tourism must be aware of the growing travel market demand from special populations and plan accordingly. (Photo of the Oregon Dunes Overlook courtesy Terry Slider)

CHAPTER 8

Techniques, Processes, and Guides

All preceding chapters have concentrated on the *what* of tourism development, especially at the community level. Information included the positive and negative aspects, demand–supply relationships, components of supply and their interdependency, special characteristics and significance of attractions, and spatial patterns of development. This chapter focuses more on the *how* of development—how it can and should be planned and managed to meet today's objectives of maximum positive impact and minimum negative impact. The chapter is presented in three parts. First, some techniques and processes that hold promise for better development of tourism are offered. These are followed by brief summaries of several community tourism guides. Finally, a proposed model for community tourism development is put forth.

TECHNIQUES AND PROCESSES

From a review of the many techniques and processes used for community tourism development, it is clear that no one approach fits all situations in all respects. However, the combination of professional designers, planners, and local citizens is more frequently used, and some examples of techniques and processes should prove useful.

Feng Shui

Because all of tourism development takes place on land, all planning and design approaches considered should attempt to provide a balance between environmental protection and land use. Today, many planners and designers are making use of the ancient Chinese philosophy and art of *feng shui*. Many design firms now incorporate geomancers (those who practice this art) into the professional team for tourist and resort development in Asia. Even many communities are guiding new tourism development according to *feng shui* principles.

Feng shui has been defined as the art of living in harmony with the land, such that one derives the greatest benefits, peace, and prosperity from being in perfect equilibrium with nature (Too 1993). It is the art of placement of things, such as in landscape architecture, architecture, and interior design. Proponents describe *feng shui* as a guide to utilize the earth's natural forces and balance *yin* and *yang* to achieve good *qi*, which renders health and vitality, success, and prosperity (Lip 1986). For site planning, *feng shui* means careful analysis and utilization of site conditions, especially hills, mountains, valleys, water resources, and winds. The Chinese belief is that all things contain spirits that have auspicious or inauspicious attributes.

According to Too, *yin* forces are considered dark; *yang,* light. These are complementary, not competitive. *Yang* symbolizes heaven, the sun, light, vigor, positive energy, maleness, and strength. *Yin* symbolizes the earth, moon, darkness, femaleness, and softness. In the landscape, *yang* refers to land forms such as hills and mountains, whereas *yin* is represented by valleys, streams, and water. Too much of either is undesirable. For example, flat land needs to be altered with some forms of elevation, and mountainous areas are complemented by valleys and lower structures and plants.

The application of spiritism to land planning in Thailand has been described in depth by landscape architect Chaiyasit Dankittikul (1993). In spite of great industrial development in recent years, Thai traditions of indwelling spirits still strongly influence land use decisions. Not only are activities planned to fit with "the geographical, topographical, and climatic conditions of the natural environment, but also with the powerful spirits of the locality" (Dankittikul 1993, 6). Designers and planners of tourism development everywhere can enhance its success by investigating the many references that describe the role between spiritism and planning to maintain the unique properties of place. Place qualities are a major part of attraction development that can be aided greatly by the application of *feng shui* principles.

An example of feng shui *applied to hotel design. The large hole in the Repulse Bay Hotel, Hong Kong, allows* chi *(dragon's breath) to flow freely from hillside to ocean, ensuring prosperity for owner, operator, and guests. (Photo courtesy of Hong Kong Tourist Association and John Ap)*

Genius Loci

Although having different origins, the concept of *genius loci,* the spirit of place, is similar in many aspects to that of *feng shui.* Both are founded in land characteristics, and both attribute more than physiological aspects to the land—a spiritual overlay. Both emphasize the uniqueness of place, that not all places are alike.

The geographer-planner Fagence (1993) has documented the many researchers and writers who have endorsed and explained the concept of *genius loci* and how important this concept can be to communities as they plan for their future. The legal foundation for zoning has merit in allocating physical space but its stereotyped application falls short of recognizing the more subtle and human dimensions of land areas. Although these human attributes of place are usually taken for granted and therefore not articulated, they are as significant and influential as physical measurable features.

Features of *genius loci* may include landscape forms evoking enclosure, building masses in contrast to open space, community entrance impressions, nodes for social exchange, environmentally sensitive areas, locations of traditional happenings, sites rich in historic or ethnic lore, and places known as special by the local population. Every community, small town, and rural area has aesthetic and spiritual values like no other. When probed, local residents can describe these values. A major component of planning and developing areas for tourism is respecting and interpreting such values for visitors.

Technologies for Sustainability

In recent years, technical advances have greatly expanded opportunities to develop facilities in a more resource-sustainable manner. An outstanding document for sustainable design, *Guiding Principles of Sustainable Design,* was prepared by the U.S. National Park Service (1993). This detailed and practical guide of 117 pages contains information on interpretation; natural and cultural resources; site and building design; energy, water, and waste management; and maintenance.

For site and building design, detailed charts identify potential natural resource threats, described in three categories—pollution, physical processes, and biological systems. Pollution includes topics such as: noise, air quality deterioration, toxic releases during construction, toxic spills, vehicle pollutants, petroleum spills, toxic or sewage releases from vessels, odor releases, and hot water discharges. Physical process threats include: erosion increases, sedimentation increases, soil disturbance or compaction, soil removal, surface water flow disruption, groundwater supply reduction or depletion, longshore dynamics alterations, and dredging. Biological threats encompass: vegetation alteration or destruction; habitat alteration, destruction, or fragmentation; coral reef damage or destruction; creation of barriers to wildlife movement; increases in roadkill numbers; invasions of exotic species; introduction of diseases; interruption of wildlife life cycles; alterations of food chains; and introduction of animal rights issues.

The guide takes a similar approach to describe threats

to cultural resources from site development. The purpose of identifying these concerns is not to stop development altogether but to alert the designer or developer to consider them in every design and development decision. Most pitfalls can be avoided with proper design techniques. All planned development should consider all potential impacts. Recommended concepts are described and illustrated for building and site design.

Perhaps the most dramatic of the new technologies are those designed for energy management. Facility design should use "daylighting," fluorescent instead of incandescent bulbs, photo cell controls, high insulation for refrigeration, air-drying laundry equipment, low-energy transportation, electrical load management, and solar technologies. Photovoltaic systems, hydroactive systems, wind systems, and use of biogas (fuel from anaerobic digest of waste) are more feasible today than in the past.

Conservation is also being fostered with new water and waste systems. Water-conserving flush toilets, spring-loaded lavatory fixtures, and special water treatment systems should be employed. Solid waste management can be improved with biodegradation systems, composting, anaerobic digestion, and recycling.

Many companies today, such as DynCorp EENSP (Buttner 1995), provide technical assistance to developers of ecotourism sites and facilities along with manufacturing products for more efficient development and management.

One caution regarding the use of new technologies (as described in Chapter 2) must be given. New technologies of self-sustaining electrical power and waste treatment may encourage developers to locate ecotourism operations in remote and fragile resource areas where human use should be prohibited. Local policies and collective local judgment should be exercised for control of this threat.

Small photovoltaic systems have much less visual and aesthetic impact than large ones. Excessive use of wind systems can have negative aesthetic impact, especially in cultural resource settings. Modern computer visual simulation techniques can help developers visualize such impacts before building (Renew 1995). Photovoltaic and other systems need not be located directly on critical areas where their appearance would be detrimental to the visitor's enriching experience.

Interpretation

Visitor interpretation is expanding rapidly as a function of attraction development and management. Visitor interpretation can be defined as the interface between visitors (their interests, expectancies, knowledge) and attractions (things to see and do). This interface is the key to whether visitors enjoy and are rewarded by their contact with things and programs as they travel.

Originally begun by the National Park Service, interpretation is a broad function now being carried out extensively in natural and cultural attractions, museums, zoos, aquariums, and festivals. Its purposes include description, guidance, understanding, and entertainment. These purposes stem from the need for visitors to know what they are viewing or doing, which directions and walks to take within a site, how to behave in a strange environment, how to gain a full understanding of the experience and to enjoy themselves. In response to this need, interpreters are being trained and interpretive visitor centers are being built and staffed in a great many destinations.

Perhaps the greatest accomplishment of the interpretive visitor center is the reduction of mass tourism impact on resources. Research by Graman and Vander Stoep (1987) has demonstrated that destructive visitor behavior can be reduced appreciably by interpretive techniques. Instead of great volumes of people being allowed to invade areas of wildlife, rare plants, archaeological sites, and historic buildings—creating noise, landscape erosion, and disturbance—they are brought into a large building where they can benefit from a vicarious experience without degrading the environment. An interpretive center should be located on a "hardened" site near the attraction nucleus rather than directly on or within it. It should be directly accessible from main travel routes. Ample parking for motorcoaches and RVs as well as passenger cars should be provided.

The center itself could contain the following features:

- an entrance lobby designed to carry out the theme of the location and to include: clerk and counter for registration, toilet facilities, and a waiting area for groups;
- an exhibit room with dioramas, displays, mazes, video presentations, and models depicting the several subthemes important to the location;
- an auditorium for audiovisual presentations and lectures;
- a shop for sales of books, photographs, videos, and souvenirs;
- a theater for pageants and other performing arts related to the site;
- a lunchroom;
- classrooms for schoolchildren and university student use; and
- staff rooms for administration, research, meetings, and preparation of exhibits.

The surrounding area could include short interpretive tours, either guided or self-guided. Often, outside exhibits and demonstrations can provide more interpretation than is given inside a center. Although in the past interpretive functions have been provided by the public sector, new opportunities exist for the private sector. When communities and destination areas begin to develop tourism, the establishment of interpretive visitor centers should accompany the development of all attractions. Equally important is proper training of professional interpreters and tour guides. An experienced interpreter of the National Park Service, William H. Sontag (1986) has stated that an interpreter's program should fulfill the following criteria: It should be fun, ego-involving, value-loaded, responsive, and pertinent.

Public Involvement

In the past the most popular form of public involvement in tourism development projects was holding hearings. Both public agencies and private investors would complete their land purchases and building plans within their own offices and then call for public hearings. Such an approach cannot truly be called public involvement; it seeks carte blanche approval after the fact. The call for a hearing may not be noticed by critical constituencies. Important groups may not appear because of date conflict. It is likely that those who do come do not truly represent the local population. And, in many nations where a command economy has been the rule, local people may not come to a hearing because they doubt that their voices have any clout. The typical public hearing is a poor form of public involvement.

If a better method of public involvement in tourism decision making is to be used, it must have characteristics that allow citizen input at all stages from the beginning of planning through construction and operation. Although this objective may complicate the process, it is essential for true public involvement.

McNulty (1994) has advocated strategically managed citizen participation. Several techniques have been recommended by specialists in this process. One involves the use of focus groups, in which a single topic is discussed by a selected group, led by a neutral professional facilitator.

A more comprehensive technique, the workshop, has been used very effectively for decades by the Cooperative Extension Service of the United States. A first step is to select an unbiased facilitator and structure participation in a day-long meeting on a selected list of topics. This step requires planning by a sponsor group familiar with tourism development issues. A clear statement of the workshop objectives is prepared, and equal representation from three or four categories of citizenry is invited, such as developers, planners, and local residents. Before the meeting, the facilitator and sponsoring organization prepare a series of specific questions to be answered by the participants. The meeting begins by division of the audience into "buzz groups" of five to ten people, each group made up of representation from each category. The several groups are then asked to select a leader and a reporter. The groups are then given five to ten minutes to discuss the question and to derive conclusions and recommendations. Then, in plenary session, designated

reporters provide their groups' answers, which are listed on flip charts by the facilitator's assistant.

As reports are presented from the various groups, redundancies are eliminated and a final list of answers is tabulated by the facilitator. A further refinement is to ask the group as a whole to give priority to all answers and recommendations. There are many advantages to this process. It prevents very vocal individuals from dominating the entire meeting, instead allowing open and free expression by every participant. By structuring the representation, it forces a balance of input by diverse interests. Those who participate feel as if they have had an opportunity to be part of the decision making. When this process is carried out in a friendly and constructive manner at a time and place best suited to comfortable, constructive, and even entertaining discourse, it can be very successful.

In 1994 several workshops of this type were held in the province of Newfoundland and Labrador, Canada (Gunn 1994). This project was sponsored jointly by a private-sector tourism organization and an arm of the national park agency of Canada. It was held in response to the demand for new economic sources following a severe decline in coastal fisheries. The meetings were attended by a diversity of local residents including fishermen; operators of restaurants, gift shops, and motels; city council members; mayors; historians; boat tour managers; teachers; and representatives of local and provincial government agencies.

Among the several discussions, the participants were asked to identify their most important social, environmental, and economic goals. Without any prompting, it was encouraging to observe how well local citizens were able to address basic questions. Presented below are some of the results obtained from the participants representing the northeast region of the province, the Bonavista Peninsula.

Social goals for tourism: Maintain our local integrity; initiate our own programs (not wait for the government); maintain past traditions (stories, songs, concerts, garden parties); involve all local social units for tourism development; and increase our ability to remain in our homeland.

Environmental goals for tourism: Protect natural and cultural resources; develop trails and interpretation for visitors; educate locals and visitors on litter control; link resource planning with tourism; improve off-season attractions; emphasize more planning and development of cultural and natural resources; protect small town qualities; and use tourism to enhance resource protection.

Economic goals for tourism: Encourage tourism entrepreneurship to replace declining fisheries; educate locally on tourism economics; improve service standards; increase the ability of young people to stay and work in tourism; increase the number of indigenous attractions; and encourage packages for longer stay for greater economic impact.

Today, every effort must be made to involve the public in tourism development planning and decision making.

Community Development in England

An approach to community development based on local volunteers rather than government has met with some success in English rural areas (Rogers 1993). Originating in 1909 with the creation of the Development Commission (now the Rural Development Commission), Rural Community Councils (RCCs) began formation in the 1920s. Later, during the Thatcher administration, the emphasis of support and action was moved dramatically from the state to the private sector, especially to local communities. (Government aid remains even though volunteer organizations have grown greatly.) Figure 8-1 illustrates the structure of rural voluntary community development, including tourism, in England. Funding still comes from the Rural Development Commission, but a national association, Action with Communities in Rural England (ACRE), was created in 1987, to provide support to the RCCs. Their functions can be summarized:

- monitoring and advocacy;
- direct support and advice;

- support for parish councils and voluntary bodies; and
- direct action (innovative pilot projects).

Critics of such volunteerism cite several concerns: that influence is thus captured by prevailing social and economic structures, and that it fosters a lack of innovation, allows government to abrogate its own responsibilities, and even strengthens dependency on government. Even so, greater cooperation among public and private groups is taking place as a result of greater local volunteerism.

Environmental Planning

It is at the planning and design stage that most of the potential ills of tourism development can be avoided. The World Bank (1991) has identified many such concerns and described how they can be reduced or eliminated. Table 8-1 identifies potential negative impacts from tourism development and mitigating measures that may be implemented. Reducing or eliminating negative effects may require some added costs: choosing more costly sites, expanding or building new water supply and waste systems, and adding to police and fire control staff and equipment. But in the long run these costs may be offset by protecting environmental assets that bring tourists, thereby extending the life of tourism investment for hotels, food services, marinas, and the like. Planning is especially important in destinations with highly seasonal trade. And in developing countries, where water and

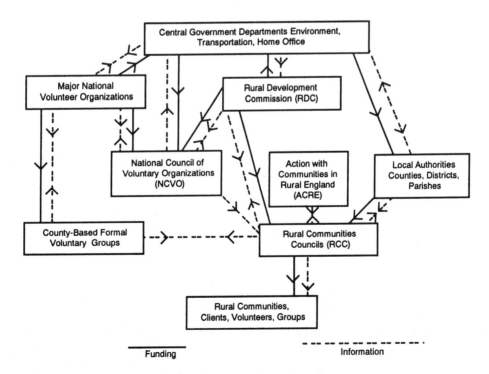

Figure 8-1. *Community planning in England. In rural areas, community tourism planning is accomplished through the program of Action with Communities in Rural England (Rogers 1993).*

TABLE 8-1. Potential Negative Impacts of Tourism and Mitigating Measures

Potential negative impacts	Mitigating measures
Direct	

1. • Beach mining of sand for construction.
 	
   • Destruction of reef for aggregate material used in construction.
	1. • Control of construction contractor.
	
	   • Submission of plans in accordance with local ordinances on beach sand mining.

2. Destruction of wetlands, forests, other unique/sensitive habitats or cultural, historical, and archaeologically important sites.	2. • Areas considered for development should have zoning plans to account for natural geographic and socioeconomic condition. • Base development phase on an inventory of resources.
3. Erosion resulting from uncontrolled clearing, infrastructure construction such as roads and marinas.	3. Develop erosion and sediment control plans.
4. Loss of "free" environmental services from natural systems and degradation of air, water, land resources.	4. • Carrying capacity should be defined so that target tourist population can be sustained without overburdening existing infrastructure and resources. • Include improvements in project design.

5. • Water pollution from inappropriate sewage or solid waste disposal.
   • marine effluent disposal
   • residential sewage disposal
   • marinas
   • infiltration to groundwater
	5. • Allowance made for use of existing municipal or regional collection and disposal system or construction of on-site sewage treatment plant.
	
	   • Liquid waste should not be discharged onto beaches, coral reefs, or other sensitive areas.
	
	   • Verify local capacity to monitor and enforce pollution regulations.

6. Solid and liquid waste disposal creates nuisance conditions adjacent to amenities.	6. • Appropriate waste disposal options required to manage potential problem. • Landfill versus incineration alternatives, as well as waste minimization will be considered.

7. • Access problems created:
   • traffic congestion
   • noise
   • minor and localized air pollution
   • people density greater than services available
	7. Access problems minimized by integrated planning to reduce traffic and pedestrian congestion, noise.

8. Sea turtle nesting affected (special case).	8. • Beach monitoring for turtle protection, coupled with beach zoning and development guidelines, to preserve the natural beach environment from the primary dune seaward. • Restricting night activities on nesting beaches during egg-laying and incubation periods.
9. Displacement of human population.	9. • Plan and implement program of compensation and resettlement. • See Chapter 3 for discussion of involuntary resettlement concerns.
Indirect	
1. Conflicts with other resource use such as fisheries, agriculture.	1. • Conceive tourism development in framework of national, regional, local socioeconomic development plans to integrate new objectives into development strategies. • Identify zones most suitable for tourism.

TABLE 8-1. Potential Negative Impacts of Tourism and Mitigating Measures (*continued*)

Potential negative impacts	Mitigating measures
Indirect (*continued*)	
2. • Stress to capacity to manage the "tourist or related environment." • legislation and polling constraints • agency support lacking • staffing and financial resources to mitigate impacts absent/reduced • inadequate training in environmental management	2. • Comprehensive legislative action frequently required to address direct and indirect impacts and their monitoring and evaluation. • Staffing and equipment support must be budgeted, including whatever training necessary to mitigate impacts and monitor the "environmental protection plan" or other mitigation plan.
3. Multiplier effect on other industries causes increased stress on natural resources or services (craft market, vendor, taxi driver, suppliers, farmers/fishermen).	3. • Provide adequate infrastructure and services support to meet physical, social, and economic needs of the region. • Recognize that "overbuilding" may be a persistent problem.
4. Congestion, overcrowding.	4. Design (urban areas and transport networks, etc.) according to carrying capacity of natural setting.
5. Natural hazards peculiar to developed site such as coastal storms, flooding, landslides, earthquakes, hurricanes, volcanos, may stress infrastructure and reduce long-term benefits.	5. Design facilities to: (a) meet best possible specifications for natural hazard amelioration; (b) take advantage of natural resources such as wetlands ability to buffer storms or absorb treated wastewater (see "Natural Hazards" section).

(*Source: Environmental Assessment Sourcebook* 1991, 228–230).

waste carrying capacities may be low, much larger capacities will be needed to cater to visitors with higher demand. Coasts and islands, because of their limited resources, demand special care in planning to avoid degradation of these resources.

Too often, only one solution to a tourism project is planned. Because of the many variables that can influence the design of a park, historic site, hotel, restaurant, tour, or other tourism project, more than one alternative should always be researched and proposed. Each alternative may be seen by local residents and sponsors in a different way. Each plan idea may have different potential social, economic, and environmental impacts. Each one may have quite different cost–return estimates.

The World Bank's guidelines should be meaningful and helpful to developers, especially if they seek funding.

Ecolodge Planning

At a seminar on ecolodge planning, Carol Adams (1995), a landscape architect, put forth a recommended "holistic site design" process for primary facilities to meet ecotourism travel demand. It consists of four major sections: understanding of the place, understanding of the market, principles of site planning and sustainable design, and monitoring and managing the effects of development.

1. *Understanding of the place.* This first step involves analyzing characteristics of the context of the proposed site—its surrounding development, infrastructure, communities, and energy sources. Features of the natural and cultural environment are identified. Potential negative and positive impacts on local communities are projected, such as jobs and environmental changes.
2. *Understanding of the market.* In addition to identifying visitor market characteristics, at this phase planners begin to create building and site program concepts—activities that will be most suitable and objectives for the ecolodge. At this point many questions are raised regarding the aesthetic atmosphere, management potential, physical arrangement, and facilities to be provided.
3. *Principles of site planning and sustainable design.* This phase begins with using local labor and energy-saving techniques. Emphases may include environmentally sensitive design, access and site circulation, energy and waste systems, and aesthetics. The design process is summarized as: programming, site inventory, soliciting community involvement, site analysis, site opportunities, design concepts, preferred alternatives, phasing plan, and government approvals.
4. *Monitoring and management system.* This final phase involves setting an environmental baseline and setting up continued monitoring of management prac-

tices in the future. Emphases are on visitor experience, community involvement, and team cooperation among designers, citizens, and managers of facilities.

Scenic Highways

As the interest in scenic roads continues to grow, so does the search for guidance for their development. Tourist market demand, especially for automobile and motorcoach travel, remains high, requiring new techniques and processes for highway planners and local tourism developers.

Psycho-physiological research has documented the fact that positive reaction to scenery has a sound scientific basis. Studies by Ulrich (1974, 1993; Ulrich and Addoms 1981) and others have tested subjects' responses to experiencing natural resource settings and scenery with sensors of heart rate, skin conductance, muscle tension, and other factors. These studies have shown restorative and stress-reducing responses from exposure to nature that, in the tourism realm, can be associated with economic value because of the great number of tourists who seek such experiences. The scenic highway is but one of many tourism applications of this theory.

Basic factors for creating scenic highways include: scenery meeting specified criteria, viewsheds, and visitor interpretation. Even casual review of these factors suggests that the development of scenic roads requires more than mere designation. Highway maps often label some stretches of road as scenic based only on the opinion of the cartographer. The roadside conditions can change so much over time that such labeling becomes obsolete.

One of the most detailed approaches to scenic highway approaches is performed by the Parkways, Historic and Scenic Roads Advisory Committee of the Arizona Department of Transportation. This process is described in *Application Procedures for Designation of Parkways, Historic and Scenic Roads in Arizona* (Arizona Department of Transportation 1993). The advisory committee is composed of eleven members—six citizens appointed by the governor, plus representatives from the Arizona Department of Transportation (ADOT), Arizona State Parks Board, Arizona Historical Society, Arizona Office of Tourism, and Tourism Advisory Council. The first step is initiation by any advocacy group requesting scenic roadway designation by the advisory committee. This committee reviews, prioritizes, and evaluates each request based on established criteria. Recommendation is then made to ADOT, and guidelines are prepared. Distinction is made between parkways, scenic roads, and historic roads.

After a road segment is identified as potentially scenic or historic, an inventory of its resources is made. For a scenic road proposal, the following factors surrounding the road are described: geology, hydrology, climate, biota, and topography. For a historic road proposal, the

factors are: architectural resources, historical resources, archaeological resources, and cultural development.

For both scenic and historic roads a *visual assessment* is required. This assessment includes the following steps:

- Identify a 15-mile zone on either side of road.
- Identify the biotic communities within this zone.
- Identify biotic zones of transition.
- Identify vegetation in each biotic community.

Then, a *landscape inventory* is made, which includes descriptions of vegetative cover, landforms, land uses, and other notable features of road segments and key viewpoints. Any viewshed needing special regulations is identified. Finally, recommendations at this stage concern: modifications to structures and signs, pruning or removal of plants, enhancement of historical markers, erosion control, pedestrian and traffic issues, compliance with zoning, key viewpoints, and vegetative restoration. After this stage, all proposals are evaluated on the basis of several criteria.

As of 1993, twenty-one scenic and historic roads had been designated. Even given the detailed examination and evaluation that went into their designation, the future of such roads is not necessarily assured. A viewshed may include many ownership jurisdictions, and the various landowners may seek changes to their properties. If these changes detract from the purpose of a road's designation, the designation may be retracted. It must be emphasized that regular monitoring of scenic, historic, and parkway roads is essential. There is danger that a road may become so popular that traffic congestion damages aesthetic appreciation. If a scenic road is also the only throughway and carries commercial vehicles, there may be conflict between truckers and tourists, especially regarding speed. As new demands are made on the lands surrounding a road—timber harvesting, mineral extraction, agriculture—the impact on its scenic and historic qualities must be assessed. Creating and maintaining historic and scenic roads to fulfill the growing travel market demand is a very complex issue.

GUIDES

Today, many agencies and organizations have recognized the need for assistance to communities and destinations so that they can develop tourism in a constructive manner. Governments, universities, consultants, and private organizations have begun to put forward their recommendations for getting the job done right. Following are some of these recommended guides.

Blank's Community Tourism Guide

A comprehensive community tourism development handbook, *The Community Tourism Industry Impera-*

tive: The Necessity, The Opportunities, Its Potential (Blank 1989) is full of sound recommendations. Although focused on building the tourism economy, it emphasizes also the local quality of life and the pitfalls as well as opportunities associated with development. The book is organized into four major parts: tourism definitions and fundamentals, how tourism functions, actions needed, and steps required for planning and implementing actions.

Especially helpful are Chapters 10 through 12, which provide practical steps required for community tourism development. Chapter 10 encompasses the needed decision-making processes, how to discover opportunities, and how to move ahead from fact-finding to action. Chapter 11 describes the necessity for leadership, collaboration among all players, and planning guidance that covers the roles of the following: attractions, activities, promotion, hospitality services, transportation, ambience and aesthetics, local tourist guidance, finance, and resource management. Chapter 12 discusses action steps, such as initiatives, goals and objectives, information gathering and analysis, concepts for development, strategies needed, implementation, and monitoring.

Minnesota Guide

A succinct guidebook, *So Your Community Wants Travel/Tourism?* (Simonson et al. 1988), based on many years of community tourism development in Minnesota, remains a fundamental for planning community tourism development.

The section titled "Community Leaders Need to Understand Travel/Tourism" provides insight into visitors, travel data, economic and social impacts, and tourists defined as business and pleasure travelers.

The next section encompasses three major components. The section "Travel Attractors Define a Destination Area" emphasizes that attractors offer a variety of things to see and do, aesthetic qualities, shopping areas, sightseeing, entertainment, outdoor recreation, business and convention facilities, opportunities to visit friends and relatives, and basic amenities. Quality hospitality services and facilities, such as lodging and food service, are essential. Also important is a local population that can truly host visitors. The second component is titled "Identify Potential Markets." Self-analysis for this component includes identifying and describing travel segments, evaluating the economic value of each segment, and selecting the appropriate segments to target. The third component places emphasis on "Linkages"—the information, communication, and transportation connections between a community and its target markets.

This section is followed by a section on the "Pros and Cons" of tourism development, which cites the economic and social advantages that need to be balanced against potential problems—social conflict, community conflict,

environmental quality deterioration, employment quality deterioration, and added pressure on public services.

A very important section follows: "How to Get Going." Recommended steps include: identify a potential leading organization, gain support from many community interests, obtain help from specialists, identify all ongoing development (attractions, resources, programs, transportation), set goals, identify a theme, analyze impacts on the environment, study market potential, and provide adequate information. Also recommended is a hospitality seminar for host training.

Tourism USA

At the U.S. federal government level, certain members of Congress voiced the need, particularly among small towns and rural areas, for guidance in developing tourism. Following several hearings and recommendations from consultants, the U.S. Travel and Tourism Administration commissioned the University of Missouri in 1977 to produce a manual to fill this need. The first edition of *Tourism USA* was published in 1978 and the third in 1991 (Weaver 1991).

This 214-page document, subtitled *Guidelines for Tourism Development,* provides comprehensive information on six topics: appraising tourism potential, planning for tourism, assessing product and market, marketing tourism, visitor services, and sources of assistance. The planning process recommended is illustrated in Figure 8-2. Throughout eight steps, this approach emphasizes citizens' input. The steps are: inventory social, political, physical, environmental factors; forecast/project trends; set goals and objectives; examine alternatives to reach goals and objectives; select preferred alternative; develop a strategy; implement the plan; and evaluate. Detailed description is provided for each of these steps. Throughout, the roles of both public and private sectors are described. Although prepared for use in the United States, this guide has been influential in improving tourism development elsewhere.

Community Tourism

Another guide that has been used widely in the United States and Canada is *Community Tourism Action Plan Manual* (Alberta Tourism 1988), revised edition. Prepared by consultants and bureaucratic specialists of the provincial tourism agency, the manual is presented in the following sections: "Introduction," "Organization," "Process," and "Appendices," followed by worksheets and a marketing guide.

The Community Services Branch of Alberta Tourism provided the impetus and guidance for communities to use this manual. As a result, of the 432 communities within the province, 310 had completed and registered their plans by the end of 1991 (Edwards 1991).

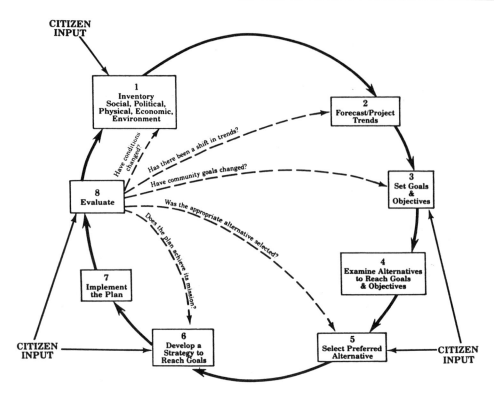

Figure 8-2. *Community tourism process. An eight-step concept for community tourism planning recommended by the U.S. Travel and Tourism Administration. It provides for citizen input throughout the process (Weaver 1991).*

However, the implementation process was more difficult. Some community leaders and volunteers experienced burnout. Funding from the province was dropped, forcing communities to seek support elsewhere. Many communities realized the single-community approach was too narrow; they therefore began to work together. The province responded by issuing the *Regional Tourism Action Plan Manual* (Alberta Tourism 1992). This manual recommends the following seventeen steps for creating tourism development at the multicommunity level: hold an introductory meeting; obtain council resolutions; create a Regional Tourism Action Plan (RTAP) committee; hold an initial RTAP workshop; identify regional assets; review market information and trends; develop regional themes; conduct a regional market analysis; identify regional tourism development concerns; set regional goals; develop regional objectives; prioritize those objectives; develop action steps; obtain community endorsement; finalize the plan; implement the plan; and monitor and review.

A positive result of this manual has been greater understanding and cooperation among the relevant communities, leading toward plans and actions for developing potential destination zones (Hill 1996). This revised approach in Alberta has demonstrated the effectiveness of shifting responsibility for developing community tourism from individual communities dependent on the government to broad alliances of local communities.

Nature Tourism Guide

Recently a community tourism development guide focused on natural resources was published in South Carolina. *Developing Naturally: An Exploratory Process for Nature-Based Community Tourism* (Potts and Marsinko 1996) provides many constructive recommendations. The six sections are titled: "Why Do We Want to Develop Nature-Based Tourism?" "Taking Our Inventory," Financing and Market Identification," "Management Techniques," "Goal Setting and Developing a Plan of Action," and "Marketing Your Nature-Based Attractions."

This practical guide offers a series of tasks to be performed: gather pertinent background information; identify key resource people; list and evaluate natural resources (water and land-based) in surrounding area; identify potential festivals, events, and recreational opportunities; identify all service businesses; set goals and identify themes; and stimulate needed development—attractions, promotion, ecotours, hospitality training.

Every community contemplating tourism development or expansion should review all guides available. Many recommendations are now available based on research and experience. In a report to the Travel and Tourism Research Association of Canada, Gordon Taylor (1994) prepared a list of available aids divided into the following categories: "A Selected List of Aids to Doing Research" and "A Selection of Aids to Using Research."

PROPOSED SOCIOECONOMIC-
ENVIRONMENTAL MODEL

A major consideration in the creation of a model of desired tourism development is the false assumption that one model can fit all situations. Every political and geographical area has a different historical background, different traditions, different ways of living, and different means of accomplishing objectives.

However, threaded through all these differences are some similarities to be observed in preparing for and achieving tourism development.

Basis of the Model

The model put forward here is in general terms, allowing a great amount of variance and adaptation to fit local conditions. Again, this model is predicated on the concept of destination as described in Chapter 6. Although the focus is on community, the surrounding context is as critical to tourism development as the features within the confines of the village, city, or metropolitan area.

Even though this model, shown in Figure 8-3, is presented here as a sequence, its true value can be realized only when it is applied in a circular, continuously rotating manner. This method is in direct contrast to the earlier approach of making and completing a plan, and then attempting implementation. Instead, the plan is never completed; it is remade regularly, offering new directions based on the results of incremental development. No developer or planner is knowledgeable enough and talented enough to anticipate all the variables of tourism that may appear in the future. Yes, there must be planning directions and guidelines; but they are temporal and must remain fluid to take advantage of the experience gained every year. Such experience can reveal potential environmental, social, and economic threats and help to resolve these issues before they grow into insurmountable barriers.

Organize

Community tourism development cannot start with only a new hotel or a new promotional campaign, because these are merely parts of the whole. If an area is to gain the comprehensive view needed for such development, the list of sponsors must be equally comprehensive. It is unlikely that any single existing organization will have the scope needed. In the United States, where it is often assumed that tourism development is the role of the local chamber of commerce, experience has demonstrated that is not often within such a group's ability to achieve. The methods of chambers of commerce for attracting industries, directed only to managers of industrial projects, do not apply to tourism. Also, because such organizations are made up of business representatives, two other very important sectors of decision makers are left out—nonprofit organizations and pertinent public-sector agencies.

A tourism development organization must include not only these three sectors of tourism developers, but also representatives of the many facets of a community and its surrounding area. These representatives must be party to ideas, discussions, and decisions regarding the community's future tourism. Especially important as natural and cultural tourism grows is incorporating all the agencies responsible for managing these resources. National, provincial, and state parks and preserves are increasingly the sources of visitor attractions.

A typical sampling of those who should be invited to participate in a tourism development organization includes, at least, representatives of: businesses, historical societies, automobile and travel services groups, museums, art councils, environmental organizations, outdoor recreation and sports organizations, convention centers, churches, educational institutions, parks, government departments, transportation organizations, financial institutions, service clubs, real estate agencies, entertainment and promotion groups, and the residential sector. Ideally, this should be a permanent organization with rotating membership. Generally, this group should represent three

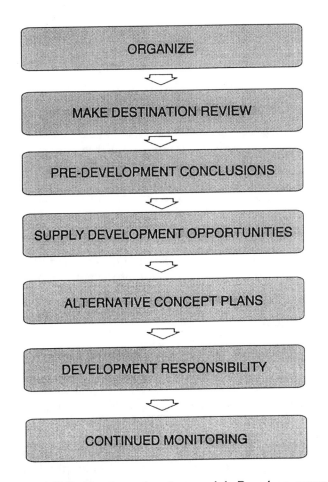

ORGANIZE

MAKE DESTINATION REVIEW

PRE-DEVELOPMENT CONCLUSIONS

SUPPLY DEVELOPMENT OPPORTUNITIES

ALTERNATIVE CONCEPT PLANS

DEVELOPMENT RESPONSIBILITY

CONTINUED MONITORING

Figure 8-3. *Tourism planning model. Based on many approaches to community and area tourism planning and development, this model is put forward by the author.*

broad categories—planners, tourism developers, and the local citizenry.

Make Destination Review

Because they have not needed to before, most communities have not studied the characteristics of their own areas, especially from the viewpoint of visitors. A publicly involved review of these characteristics can be enjoyable as well as enlightening. This step can benefit from professional planner input, but not necessarily in the traditional consultant manner.

The tourism development organization can accomplish this step most effectively by assigning the several topical investigations to task forces. At least six topics need to be thoroughly understood by everyone if tourism is to move forward: existing tourism development, travel markets and trends, natural and cultural resource foundations, potential social impacts, potential environmental impacts, and potential economic impacts. Most communities and adjacent areas contain residents with interest and often knowledge on one or more of these topics. With only small amounts of professional input, task forces can study these topics, prepare reports, and publicize their findings. Experience with this action has produced startling "discoveries" by even long-time residents. Following are suggested activities for these six ad hoc task forces.

Examine existing tourism development

Because communities vary so much in their experience with tourism, it is necessary to discover where they stand. Examination of each component of the supply side is needed—attractions, services, transportation, information, and promotion. By means of observation, literature review, and interviews, important features can be determined—location, extent of development issues, degree of success, and prospects for the future.

Key questions to be answered by this task force include the following. What can visitors to the destination now see and do? What resource foundations are these attractions based upon? What sectors own and operate them? Do they appear to be well managed? Is there potential for expansion? Is transportation to the destination adequate, and what modes are used? Is pedestrian circulation adequate? Where do motorcoaches, RVs, and automobiles park, and is it adequate? Where can visitors obtain food, lodging, car service, and shopping? Do they meet visitor market standards of quality? How do visitors find their way about the destination—by signs, tour guides, maps, radio information? What are the means for promotion?

Analyze travel markets

This task force can gather its information from two main sources. The public tourism agency should be able

to supply generalized travel market information. Market surveys often provide data on origin, length of stay, expenditures, accommodations used, and attractions visited. A second source of information is the task force's observation, interviews with lodging managers, and original localized surveys. From these sources the task force may be able to profile existing travel markets and obtain clues to markets not yet reached.

Study natural and cultural resource foundations

Because there is a tendency for local residents to take the land resources for granted, for tourism it is necessary to document these factors, viewing them through the eyes of visitors. A combination of touring the area and making a literature search can provide a task force with an entirely new understanding of these resources and their opportunities for development. If a consultant is engaged for this task, the study should make use of all local talent possible.

One factor to be studied is the natural resources and their relevance to new visitor opportunities if developed. Key factors include: land relief (hills, mountains, valleys), surface water (streams, rivers, lakes, waterfalls), wildlife (game and non-game), vegetative cover (forests, wildflowers), and climate (assets, liabilities). The extent to which these contribute to scenery and aesthetics should be identified. Questions to be asked: What is the abundance of these resources? What is their quality and what are existing threats to their quality? Where are they located?

A similar review should be made of cultural resources. These would include: prehistorical activity (races, activities, digs), historical development (settlement, events, sites, buildings), institutions (churches, colleges, universities, medical centers) whose potential for visitors has not been explored, and opportunities for expanding performing arts and special ethnic interests. Important is the quality and geographical distribution of these resources.

At the same time resources are documented because of their opportunities for new attractions, additional effort must be directed toward examination of opportunities for better transportation, visitor information, and promotion.

Study social impacts

A task force should be given the assignment of determining existing social impacts of tourism, both good and bad. Questions to be asked include the following. Are there conflicts between the host and the visiting population? Do local residents believe they are enriched by the tourists? Does tourism improve or degrade the lifestyle of residents? What might be the social impact if the number of visitors were doubled or increased even more dramatically?

Study environmental impacts

The responsibility of this task force is to examine carefully the extent to which present tourism is enhancing or degrading natural and cultural resources. Are endemic land resources such as soils, plant life, wildlife, and waters being damaged, or possibly improved, by industry, housing, and other activities as well as by tourism? Are cultural resources, such as archaeological sites, historic sites, crafts, customs, and ethnic values being downgraded or improved by tourism? Are changes needed on issues related to tourism and the environment?

Study economic impacts

A task force should describe the existing economic aspects of tourism. If data are not available from state, provincial, or national sources, it may be necessary for this group to make its own estimates. Positive information should include traveler expenditures, jobs created, and tax revenues generated. Negative information should include the costs of tourism—growth of community infrastructure, new investment, and increased promotional and informational costs. Projections of both negative and positive economic impacts should be estimated for modest or great growth of tourism in the future.

Predevelopment Conclusions

In a plenary session of the tourism development organization, written and oral reports from the several task forces should be presented and discussed. Probably for the first time, the community will now know more than it ever did about its tourism to date. The results of these reports and discussions should be publicized throughout the area. This objective analysis should provide the foundation for deriving the following conclusions.

Identify the desirability of expanded tourism

It is at this point that the community should decide whether it really wants more tourism and why. Public forums on this topic should be enlightening rather than leading toward polarized opinion on tourism expansion. It is at this stage that examples of well planned and controlled tourism development should be described; the desirable alternative of guiding tourism expansion in the most positive directions should be put forward.

Identify needed land use controls

Because tourism expansion involves greater use of land, several conclusions on land use controls may be derived. For example, if natural scenic areas and vistas are cited as very important to the local population, it may be desirable to identify these sites as needing resource

protection rather than structural development. The task force review should have revealed such information. It is likely that this resource protection will in turn become a major attraction for visitors, who also can be enriched by viewing and photographing the sites. At the same time, if tourism growth is desired, local officials can designate the land areas best suited to development of tourist services and facilities—where there are the best business possibilities and most efficient access and infrastructure servicing. Cultural as well as natural resource areas with tourism potential may need to be placed in the public trust so that they have long-range value for the overall tourism system as well as for local quality of life.

Identify needed social measures

The task force study of potential social impacts of tourism's growth should have identified measures to enhance positive social exchange and reduce negative impacts. Although at this stage it is too early to finalize decisions, some measures may be put forward. For example, it may be recommended that one-to-one hosting exchange be initiated, whereby local homes are made available to visitors. Hospitality training may be offered. It may be suggested that all promotional literature carry tips for traveler etiquette to reduce social friction. It is at this point that local populations need to clarify their own quality-of-life priorities and consider measures that can accept visitors and yet protect those priorities.

Identify needed economic measures

The task force activity should have provided overall information on existing tourism economics. This effort should have raised many questions about future growth: If tourism activity is multiplied many times, what may be the result in jobs, incomes, and tax revenues? On the other hand, for tourism to grow, where will new investment funds come from; what new public funds will be required for expanded water supply, sewage handling, police and fire protection, and street improvement? It is too early in the planning process to provide solutions, but the issues must be raised and measures offered to address them.

Identify market–supply balance

The task force studies should have alerted everyone to both voids and opportunities for tourism development. If, for example, it is found that a strong travel market demand exists at an origin for river rafting and adventure, there may be opportunity for new attraction development because the resources have not yet been tapped. A comparison between market preferences of the day and what the destination now offers may show many gaps. Some new developments may be feasible, whereas others may not have the needed resource foundations. Gaps in information, promotion, and transportation may also be revealed at this stage.

Supply Development Opportunities

At this stage, a sufficient understanding of tourism's complexity and its relation to the special local characteristics allows the beginning of a new vision. Because the tourism organization is made up of local representatives, all challenges and opportunities can be visualized through the lens of local rather than outside influences. Because professional designers and planners and other catalysts have attended to local tourism interests, the analysis and conclusions from the previous steps are of high quality and full depth. Because both resources and market trends have been studied, opportunities can now be revealed for striking a new balance between market and supply. Ad hoc committees can now be assigned to identify opportunities, specifically adapted to local conditions, for all five components of the supply side.

New attraction opportunities

Studies of the land and cultural resources should by now have resulted in many clues for establishing new opportunities for visitor activities and experiences. After these opportunities have been identified, together with their potential market segments, they need to be brought before the three sectors of developers—public, nonprofit, and commercial enterprise. This publicity should arouse new interest in how these opportunities can be translated into reality. At the same time, recommendations need to be made on how and where the potential attractions can be established in the most sustainable manner.

Transportation opportunities

Problems and needed changes for enhancing vehicular and pedestrian movement should now be evident and viewed as development opportunities. Probably no other urban issue is of greater concern worldwide than traffic congestion. In many locations it is already so serious as to dampen tourist market potential in spite of superior attractions with market demand. All transportation officials and planners should become partners with tourism interests in offering solutions.

Service business opportunities

Following the identification of new attraction and transport opportunities, the need for improvements in existing services and demand for new services should also become evident. These conclusions, when brought before hotel, restaurant, and shop investors, should readily stimulate new offerings adapted to local need and conditions.

Information opportunities

When more local residents, through participation in tourism analysis, have come to see their communities through the eyes of travelers, there undoubtedly should emerge many recommendations for better tourist information. The need for improved way-finding and descriptive information in guidebooks, maps, and other literature should now become clear.

Promotional opportunities

The examination of the extent and effectiveness of existing promotional programs should have laid the foundation for new organization, methods, and product development.

Alternative Concept Plans

Based on all information and opinions gathered, the local tourism organization, with input from designers, planners, and potential investors, can now create alternative guidelines and concepts for what should (and should not) be done. Such concepts can be expressed in two categories—physical and program plans. These plans grow out of not only the needs that have been identified but also the knowledge of their potential social, environmental, and economic impacts.

Because of the complexity and capricious nature of tourism, these concepts need to be very flexible. They need to allow for changes in national and local policies and economies as well as shifts in travel market demand. And, certainly, they must be created alongside all other physical and program plans for the communities and overall destination area.

An important element of these concepts is staging. Too often, all a destination's short- and long-range needs and opportunities are presented at once, with little probability that they will be addressed as planned. It is well to have the overall concept expressed as a guide, but all plans must be presented for application in stages. High priority should be given to the most readily feasible projects. Incremental development that allows small successes at first provides the foundation for continued public support for larger and longer-range development.

Development Responsibility

Throughout this process, publicity should be sent out at each step so that local and area awareness is achieved. Such publicity should identify each of the concept plans and the many recommended projects that have resulted from the earlier stages of study and deliberation. Each concept should be explained in detail, including the justification for the projects and actions recommended.

The main focus of this step is to alert potential developers. Each of the developer sectors—government, non-profit organizations, and commercial enterprise—is made up of many individual developers. Each has its own policies and practices for making developments and may see different opportunities for its own self-interest.

For example, a concept plan may recommend the protection of a land area that could become an important attraction because of its outstanding natural scenery. Members of each of the three sectors can determine whether such a project is within their scope. Is it most likely within the policies and traditions of the local or provincial park agency, or is it a better fit for a nonprofit conservation group or even a profit-making firm?

It is at this stage that each recommended project should be reviewed by the several prospective developers for accepting responsibility. By means of publicity through workshop meetings, publication of reports, public hearings, and even direct contact with developers, a willing investor may be found. It is only at this point that project feasibility can be determined. Even within the commercial sector, each investor has his own policies regarding business and profit expectations.

Continued Monitoring

The main thesis of this model is its dynamic flow, its concept of tourism development as a never-ending process. Although it is presented here in a traditional planning sequence, the use of this model does not stop when a project report is finished, as if at the end of a consultant's contract. The tourism organization continues, the planning process continues, and the citizen involvement in tourism continues.

It is essential for communities to understand that as every increment of tourism is completed it changes the balance of the tourism system. Because this balance is not completely predictable, even with good planning, the success and impacts of each operation need to be monitored. How each affects the environment, the economy, the society, and parts of the other supply components is relevant to every subsequent development. How other community development impinges on tourism is critical to the implementation of other plans. It is likely that major changes throughout a community and its surrounding area will take place in five, ten, and twenty years, and beyond. Thorough and continual monitoring, and making recommendations

for best adjustment, are essential to the operation and success of this or any other tourism development model.

CONCLUSIONS

Although all preceding information in this book has provided building blocks for constructing worthwhile tourism development, the book's value will lie in the planning and action that is applied in the field. The brief descriptions of sample approaches to destination and community development of tourism may appear to be so different that they have little in common to offer those seeking tourism expansion. However, these and other experiences in planning and development suggest similar important inferences. All seek expansion of tourism, many seek economic gains, and generally most desire to avoid the negative impacts of the past. Recent guides to development recognize the great complexity of tourism—and that areawide and communitywide cooperation and support are essential.

Compared with past approaches, two major new trends have appeared. First, there is a new awareness of grass-roots involvement rather than top-down governmental decision making. At the same time that communities are increasingly seeking the economic benefits of tourism, they are also demanding a stronger role in decisions that affect them. Communities are willing to accept assistance but are becoming increasingly aware of pitfalls if tourism is not developed properly. Second, the process followed by professional planners and designers has shifted dramatically, from exclusive in-house plan preparation to public involvement throughout the process. Professional training in the past led planners and designers to believe they knew best what communities should do and have. It is now recognized that this egocentric approach resulted in many reports and plans remaining on the shelf collecting dust rather than being implemented. Unless the principal actors are involved at the very start, there is little likelihood of plans ever being enacted. Professional input is still needed but is now expressed in different ways.

It can be concluded that tourism is a major social and economic force worldwide. It cannot be escaped. For it to provide the goals and objectives desired by communities and destinations, there are approaches to be found that are productive and also avoid tourism's potential negative impacts.

CHAPTER 9

Conclusions and Principles

The purpose of this section is not to summarize the book, but to draw conclusions that may be helpful to planners and developers. For ease of understanding, these conclusions are grouped into the following categories: fundamentals, pitfalls and caveats, responsibilities, methods of planning and development, and design principles.

FUNDAMENTALS

No matter where in the world tourism is developed, similar fundamentals apply. When these basics are understood and applied, many of tourism's ills can be avoided and greater success can be assured. Even though physical development and programs must be adapted to varying local conditions, certain elements are encountered everywhere.

Balancing the dynamics of demand and supply is a critical challenge everywhere. What the travel market seeks and what sort of physical development is suited to local resources are in constant flux. Thus the prevailing notion that the only route to tourism success is via greater expenditure on promotion is not only naive but also a likely waste of funds. On the other hand, when tourism is treated as a system of interlocking parts, progress toward desired goals is much more feasible.

It may be concluded that the overall functioning of tourism can be modeled as a demand (travel markets) and supply (physical and program development) system. When communities, destinations, and nations view tourism in this holistic manner, its true complexity as well as its many opportunities are more readily seen. Essential to this vision is the reality of the interdependency of every action and actor within this system. Within the supply side are key components such as attractions, transportation, services, information, and promotion.

Every community anticipating tourism growth must research answers to potential traveler segments, their origins, and their interests. Only then is it possible for the community to direct programs and physical development of the supply side to meet market demand.

There is no point in investing in new destination development unless the transportation trends are clearly defined. There must be a likelihood of reliable, comfortable, and reasonably priced access from tourist origins. This fundamental requires that leaders of local tourism development obtain cooperation and clear information from transportation suppliers. In spite of great resource potential and enthusiasm, tourism does not have much chance for success unless visitors have good access.

Key to all development are the attractions, dependent largely on natural and cultural resources. As the most powerful component of the tourism system, attractions require extreme care in their design, development, and management. To draw and satisfy visitors, all attractions must be worthwhile. It is not enough for local people to take pride in their community's features; they must also obtain insight into travelers' interests. Especially valuable as an edge against competition is to emphasize indigenous attractions.

In a market economy, once access and attractions are in place, the demand for new services and facilities becomes clear to potential investors. Sensitivity to a wide range of market interest, from deluxe to economy, is essential for establishing lodging, food service, automobile service, travel service, and other commercial enterprises.

Perhaps the weakest link in the tourism system today is traveler information—on transportation, attractions, and services. Telecommunication offers great opportunity today, in addition to the traditional sources of guidebooks, maps, interpretative visitor centers, and tour guides.

Probably the most solidly entrenched component of tourism is promotion. Although its value cannot be questioned, there is logic in responding to today's needs by diverting many of the funds now used for promotion into research, planning, and community guidance.

It can be concluded that tourism can provide jobs and other rewards mainly when it is planned and developed as a functioning system that balances demand with supply.

RESERVATIONS

No book such as this would be complete without noting that tourism does have pitfalls. It cannot provide all anticipated positive results unless potential negative impacts are understood and avoided. Only in recent years have these been identified as belonging to societal, economic, environmental, and even managerial categories.

Cultural clash between visitors and local residents can be serious. Visitors to a destination bring all their cultural beliefs, habits, and behaviors—which are often in direct contrast to those of the local residents. Market selection, training, and education can be used to avoid many societal issues, but certainly destinations must understand these potential dangers before they launch heavy tourism development programs.

Recently, it has been documented that tourism's economic advantages come with costs. New jobs, sources of income, and investment wealth are often accompanied by new outlays of funds for water supply, waste disposal, police and fire protection, and the development itself. A realistic view of tourism must include estimates of these costs to determine that it can bring a favorable economic balance.

The erosive impact of tourism on cultural and natural resources is now an unquestioned fact. Tourism does use land and resources. It can consume the very assets that are the foundation for local attractions—the basis for tourism itself. Environmentalists, planners, and other specialists can identify environmental dangers if involved early enough in the planning process. Development can then be created in ways that avoid water, air, and noise pollution; soil erosion; destruction of native wildlife and plants; and permanent extermination of valuable cultural assets.

It is important to comprehend the reality of potential negative impacts from tourism and the cardinal need for planning to avoid them.

RESPONSIBILITIES

Generally, "tourism development" implies only the private commercial sector. Certainly this sector carries out a very important role throughout the world. Even countries that have formerly espoused a socialist system are finding the value of privatizing tourism functions. Generally, tourist businesses have increased their standards of service and management in recent decades. Today's trend is the acceptance of new ethical practices. Instead of waiting for governments to impose new regulations, many businesses are now practicing codes of ethics that apply to guest behavior as well as business practices. Especially significant is business acceptance of environmental responsibility. The role of the free enterprise sector is very important in all tourism planning and development.

Worldwide, governments and their various agencies continue to make many decisions on tourism's physical and program developments. Most nations, provinces, and states have forest, park, wildlife, conservation, water, and economic development agencies. Their policies often support but sometimes restrict tourism development. However, most frequently, policies and decisions are not at all focused on tourism and visitor interests. Financial incentives, for example, are most often directed toward manufacturing and industry, not tourism. At the local level, as communities gain a better understanding of tourism, their town and area councils can perform a valuable role in guiding land use to protect resource assets important to residents and tourists as well as to foster tourism businesses where they can succeed. Future tourism growth is highly dependent upon the functions of governments at all levels.

The role of the nonprofit (voluntary) sector is increasingly significant in all tourism development. The many associations focused on history, culture, politics, business, recreation, parks, and conservation are having a greater impact on tourism than ever before. In many instances they actually own property, restore buildings, protect wildlife, promote better water and air quality, and even build hotels and other tourist services. Communities can improve their tourism development substantially by making sure nonprofit organizations are involved.

Finally, the role of consultants must be considered for all tourism planning and development. Consultants with important roles in tourism include architects, landscape architects, archaeologists, historians, marketers, and professional management firms. Even though they can provide critical input, too often they lack sufficient understanding of tourism. As a consequence, the wealth of their talent, training, and experience is often misdirected. Planners and designers, for example, have great opportunities for improving the quality of all tourism development; but their traditional processes may need to be modified so as to ensure greater public involvement and greater understanding of travel markets and thus be most effective.

It must be concluded that cooperation among all sectors and all their members is essential if tourism development is to take place in the best manner. This is a desirable principle that today requires new mechanisms to bring the several actors together to share roles and responsibilities.

METHODS

From any review of tourism development, it becomes clear that no single approach fits all situations. Especially at the community level, several factors may vary greatly—tradition, policies, regulations, location, and physical orientation, to name a few. Every tourism development approach, therefore, must adapt to these local conditions. As was stated in the beginning of this book, communities have the options of refusal, haphazard development, or planned and incremental development.

Even so, there are some general basic steps that apply everywhere. In all locations, for best tourism there must be fact-finding, public involvement, goal and objective setting, action by the several sectors, and monitoring.

Too often, tourism is approached with little fact-finding. Collecting information on travel markets, especially the various segments and recent trends, is essential. Equally important is gathering data on the several factors that will influence development locally. Among these factors are government policies, organizations, and leadership; prevalence of entrepreneurship; availability of finance and labor; and particularly natural and cultural resources. These facts, together with present development, provide a substantive foundation for further planning and action.

The traditional plan–implement sequence, practiced by tourism planners for many years, is now being supplemented by a new paradigm—public involvement. Professional planners and designers are still essential, but all of their steps must have input from those to be affected by the proposed development. This revised method is probably the greatest innovation in tourism planning today.

Each community and destination should clearly identify its goals and objectives. Goals can be classed as social, economic, and environmental. In contrast to goals (aims, aspirations, desires), objectives are specific projects to be accomplished by a target date. A tourism development council can establish goals and objectives for both short- and long-range guidelines. It is essential that these guidelines remain fluid so that incremental changes can be made over time.

Key to tourism development everywhere is the action taken by the three sectors—private enterprise, nonprofit organizations, and government. All sectors' physical and program actions should be taken in context with markets, and their goals and objectives should be set locally as well as at the destination and national policy levels.

Monitoring is frequently missing from development processes. Annual reviews of progress can reveal how successful previous actions have been, as well as the issues and obstacles that need attention. This may be the most important function of an umbrella organization such as a tourism development council.

No matter which method is used, tourism development today must strive to avoid the negative social, economic, and environmental outcomes of the past.

DESIGN PRINCIPLES

The conclusion to tourism development occurs when physical projects are designed and built on the land. This is the realm of architects and landscape architects. In concert with all who are party to final decisions, espe-

cially developers, the shaping of the land, the layout of facilities, and the design of structures and the landscape are performed by these professionals.

Tourism, by its very nature, tends to make all the world look the same. There is homogeneity in the chains and franchises proliferated by the business sector. Although seeking the unique characteristics of different areas, tourists tend to reduce the individuality of place by carrying their cultural baggage with them wherever they go. This is the paradox of tourism and therefore its design challenge.

Many people have abdicated their relationship with land because the modern workplace confines them to an artificial and urbanized environment, denying them the contact with the land that was vital to their ancestors. Perhaps it is only through travel, especially pleasure travel, that they can recover the fundamental tie between themselves and place. Upon the designer's shoulders rests the responsibility of defining the essence of place while satisfying the interests and wishes of a continuing flow of outsiders. Landscape architect Garrett Ekbo (1969) expressed this concern:

> Will the proprietors and promoters of environments and facilities for tourism, those operators of obsolescent economic and technical structures, those eager opportunists in a modern no-man's land, recognize the possibility for building a program which could be of major aid to modern man in his search for a way out of our contemporary computerized dilemma? . . . Human imagination and creativity have transcended many staggering challenges on the long road from cave to skyscraper. We can hope that they will also conquer this latest and largest problem.

Certainly, the answer lies in imagination and creativity—the vision of a better use of all resources to offer a more human, richer, and more meaningful travel world. "The designer must be able to represent needs and tendencies of the client and the user to himself in order to formulate the design problem. Likewise, he must be able to represent his own intentions to the client in order to develop viable design solutions" (Akin and Weinel 1982).

But even when more creative tourism design can be agreed upon as an objective, designers and developers must seek better directions. Even beyond facts and concepts for better design, can the complicated phenomenon of tourism be given some design structure? The following ten principles of tourism design are challenges to designers and developers alike. When they are applied to projects of the supply side, everyone will benefit—visitors, developers of all three sectors, and residents of destination areas.

Functional Design

At the outset it should be well understood that all basic design principles established for all design professions apply as well to tourism land use as to any other. Those who create the tourism environment utilize all traditional and contemporary design approaches. A multitude of decisions—for example, those regarding the alignment of drives and walks, the positioning of buildings, and the development of overlooks, as well as building exteriors and interiors—must take into account all the sense perceptions of the user in addition to the characteristics of the resources. To state it broadly, the tourism environment must function, and three different functions be satisfied.

Elementary to all design, especially for tourism, is *structural functionalism*. All structures and landscapes must withstand the bombardment of inside and outside forces. They must endure exposure to weather, the stresses of mass use, and general wear and tear. In other words, all elements must be structurally sound. So much is known today about soils and the properties of building and paving materials that there is little excuse for poor structural design. The safety of visitors and the long-range durability of capital investment in tourism development demand enduring structural design.

Equally important is *physical functionalism*. The tourism environment must accommodate the many activities and movements of people and vehicles. A quantifiable amount of space is necessary, for example, to park a given number of cars, to provide food for a given number of people, or to enable a given number of people to camp. Although the limits vary, rough spatial standards have been set for a great many kinds of sports, games, and recreation areas. So-called oversaturation of destinations and deterioration of environments is not caused as much by large numbers of visitors as by inadequate design and poor management of crowds. Masses of visitors can be handled safely and comfortably through design that is physically functional.

But an environment may be sound structurally, serve its physical functions well, and yet fail the users. The most critical functional goal—and the most difficult to achieve—is *cultural* or *aesthetic functionalism*. "The structure of the building can be explained and the strength of materials tested, but the spirit of the building, its form and spaces, lines, textures, openings, and solids must be felt in much the same way the ancients sensed spirits within the forms of rocks and trees" (Wilson 1984). Environments must provide the values and images that travelers associate with development. Spatial relationships, balance, sequence, repetition, proportion, scale, color, atmospheric perspective, and all other principles applying to the arts are equally valid in the tourism environment. Most important are the fundamentals of unity and composition.

Close cooperation between developers and designers can dramatically raise the present levels of performance of each of these three functions within the tourism environment.

Sites, Buildings, and Spaces

Tourism demands more collaboration among design professionals than do most other land uses. Architects, engineers, landscape architects, interior designers, and sculptors of landscape amenities must pool their talents to produce successful tourism environments.

Buildings provide not only internal functions but also spatial relationships in the landscape. The use of these spaces by visitors is probably more critical than the use and impressions of residents. Visitors bring greater expectations and spend more time viewing scenes and taking pictures. Building shapes, heights, edges, configurations, textures, and styles are an integral part of the entire landscape. Too often design considerations of buildings are limited to two-dimensional internal functions.

Site spaces are not merely the lands unoccupied by buildings. Realtors often consider empty site space as wasted, when actually it may be very important to building functions. Designers should study how visitors may relate buildings to sites before they complete their designs of complexes. Spatial and building design is especially important at locations where long vistas are essential to tourism. Otherwise, scenic views of mountains, beaches, and landmarks may be blocked forever by a lack of sensitivity to building and site relations. The materials of the landscape architect—landforms, plants, pavement, site structures, water, and buildings—can be used effectively in protecting basic land assets, satisfying visitors, and stimulating economic success. But such integrated design demands early collaboration by all parties.

Clustering

Although Chapter 5 has already presented much about complexes for the clustering of attractions, the principle needs further emphasis as an element in total land design. Clustering is virtually being forced on developers by mass travel, which fosters the grouping of attractions, facilities, and services together. For attractions, area and site clusters differ.

The *area cluster* is a grouping of several sites by theme, promotion, and transportation linkage. The so-called heritage trail is one example; the massive development of several lakeshore subdivisions is another. The creation of a new park or recreation area often forms an area cluster. Attractions on several sites can be linked to form a cluster from the visitor's perspective. The success of the design of such a cluster depends on how well it is linked, both physically, on the ground, and mentally, by the user. The package tour is one very effective medium for creating an area cluster. By the same token, however, all elements in a cluster, even those separated by great distances, should have some continuity of design.

The *site cluster* is borrowed from the shopping center concept. Here, for example, one building mass along a beach may include a variety of water-oriented recreation. Generally the site cluster has one parking area, and the attractions are to be reached and enjoyed on foot. Decision makers and designers today are creating new microenvironments for such clusters. The pedestrian mall has been abused and misused as a tool for accomplishing this integration. But more and more settings foster walking because pedestrian development is more fun for the user, generates greater sales for business, and conserves other lands by concentrating and localizing intensive functions.

The *service–business* cluster, also exemplified by the shopping center, has many applications to tourism. Long ago, proprietors of motels discovered that their success depended less on highway orientation than on nearby destination objectives, entertainment places, and food services. The isolated motel operator generally feels the pinch of competition from "motel row." Likewise, in the tourist entertainment field, the "strip" is an organic expression of the cluster theory. The concept of a service–business cluster for tourism is suited to urban areas as well as to small towns and rural areas. A traveler is more likely to seek the greater diversity and volume of services when they are located together. And businesses in such clusters benefit from local as well as travel trade.

Recently, the *transportation cluster* (intermodal travel) has begun to appear in response to the need for better travel continuity. Seldom in the past were expressway interchanges, bus depots, airports, and train or mass-transit terminals interconnected. Again, multiple owners and myopic business visions prevented an integrated transportation system. A transportation cluster, designed to make travelers feel welcome, can set the tone of the travel experience thereafter.

Clustering has many advantages. From the developer's perspective, it is more efficient to operate, better suited to visitors, and easier to manage. It protects environmentally sensitive areas while concentrating mass use on less sensitive ones. Finally, clustering offers promotional advantages; larger complexes can have greater impact when promoted in market areas.

Suitability

Probably no stronger arguments have been made by tourism purists than those on the topic of appropriateness. Wilderness buffs have heatedly argued that automobile and airplane noises, and the visual evidence of man's control over nature, have no place in the wilderness. Supporters of national parks deplore their growing commercialism at the same time they patronize such business services. In the minds of the users and decision makers, suitability is not a uniform concept.

What is appropriate? Many would say that the popular Mt. Rushmore memorial is a defacement of natural resources, that enormous sculptures of four U. S. presidents have little or no relationship to the rocky South Dakota landscape. In this age of creating attractions, those with little skill or knowledge of appropriateness often thrust upon the traveling public a mass of inappropriate items, while treasures go unnoticed. Well-meaning citizens, hoping to cash in on the supposed affluent tourist, sometimes go to ridiculous extremes to spend hard-earned money on creating pointless attractions.

Certainly, appropriateness is a cultural concept, and extremely difficult to define; but so is love, and no one would deny its existence. If a musical pageant that uses colored lighting on a natural waterfall as a backdrop helps to explain the historic and cultural significance of that waterfall, and if the visiting public is thrilled by the spectacle, there should be little complaint about it. However, if the settings, sounds, and events surrounding the waterfall are garish and distracting, they are out of place. Much depends on the quality of the offering, which is the designer's responsibility. The designer can be innovative, but his creations cannot go far beyond current tastes, traditions, or customs.

Many aspects of the problem of suitability can be eased by carrying on the land research and synthesis required to gain a full understanding of development potential. Landscape architect J. O. Simonds (1985) has astutely pointed out the interdependence between land forms and the structures that rest on them:

> Let us consider a summer weekend vacation lodge. If this lodge were to be built on a sheltered rock-rimmed inland lake in northern Maine, its abstract design form would vary greatly from the form it would have if located anywhere along the wind-whipped coast of Monterey, California, or in the smoky Ozark Mountains, or on Florida's shell-strewn Captiva Island, or along the lazily winding Mississinewa River in central Indiana.

Designers and developers have the opportunity and responsibility of creating land and building designs most suitable for visitor interest and indigenous place qualities.

Exposed Functionality

In today's landscape and architectural idioms, functions are becoming less and less clear. Schools look like factories, which look like office buildings, which look like power plants. Architectonic landscape design is applied willy-nilly to parks, museums, amusements, and hotels.

Somehow the subtleties of function should strike through well enough that the user is not shocked by what he finds. As Simonds (1985) has stated:

> Woe be to the designer who, by plan sequence, induces in the observer a mood or expectation not in keeping with the functions of the plan. In contrast, how superbly

effective is that sequential order of spaces and form that develops and accentuates an induced response in consonance with the preconceived experience.

Designers of the vacationscape should find theirs to be an easier task than designers of many other functional uses, because, usually, visitors anticipate the objective long before they arrive. They visualize themselves already at a convention center, resort lodge, marina, campground, historic site, or mountain top. The designer has an opportunity to play into these preconceptions and allow approaches, settings, and exterior appearances to support appropriate activities at the destination. This objective does not require a blatant and skeletal display of functions; there can be restraint and even suspense. The principle here is that the visitor's impressions from the first roadside sign to the building entrance are those that set the pace for their experience of the main purpose and function of the establishment.

Efficiency in the Experience

Today, all travelers are placing greater and greater emphasis on time, cost, and quality—the consumer commodity and service approach. More and more, tourists want to get their money's worth. Although they may appear to be spendthrifts on vacation, they still operate within an overall budget and value system.

Tourists occasionally believe that some attractions are "tourist traps." This criticism is based not necessarily on actual fraud or deceit, but merely on the belief that an attraction takes up too much time or costs too much to see or do. The grave of a local historic figure, for example, may be touted as a great attraction, but in the eyes of the outsider it becomes a tourist trap when he travels many miles over a dusty road to see something that he considers of minor significance. People are willing to spend time and money only if value is received.

Basically, this is not only a matter of distance from prime access routes; it is a matter of balance between the effort of getting there (time, money, convenience) and the quality of the attraction. In many instances the design and performance of the attraction can overcome great obstacles.

Sequence and Satiety

Because this chapter has not yet examined overall development from the visitor's point of view, it has failed to touch on one very important aspect of his makeup—his threshold of satiation. Promoters, advertisers, business managers, conservationists, and nature lovers frequently forget that sometimes enough is enough. Planners, designers, and developers who work with physical land development also occasionally forget the user's attention span.

As one views a beach and the hundreds of people on it, one is inclined to think of this activity as continuing for a long period of time. The naturalist loves to observe plants and wildlife, and his absorption seemingly has no time limit. But the interest spans of most travelers are relatively short. Observe the family vacationing at a resort: Seldom does the beach hold attention for more than one hour. The teenager, for example, bounces from it to the speedboat to the sailboat to the nearest hamburger hangout, and so on. And the teenager is not alone; the modern tourist is constantly on the move.

Levels of satiety are reached in all activities—some earlier than others. Recognition of this fact must influence the design of communities, destinations, and regions for travel. The complete sequence of daily actions of the users should be considered. For example, a wooded area encountered soon after a visit to a city is refreshing. A similar wooded area encountered after one has traveled for many hours through other wooded areas loses its appeal.

Writers and composers have always recognized that without variations in tone or dramatic impact, a piece becomes dull and monotonous. A prelude is needed to introduce the main theme of the composition, where the purpose of the action or the structure of the composition becomes more clear. Then the tempo gradually increases, or the drama is intensified, until the climax is reached. Although the composition may stop at this point, its impact may, however, be too strong without a slowing-down period.

This principle also applies to development for tourism. The psychology of set perception is important, for if an individual has been exposed to several repeats that call for the same mental and physical response, he tires. Translated into the terms of land design, this concept requires that the designer be aware of sequence. As illustrated by Figure 9-1, a continuous high level of interest (A), although desired by promoters and advertisers, is very unlikely to be sustained. Likewise, constant exposure to the same attraction (C) causes interest to wane. Peaks of interest (B) spaced with low-interest or contrasting activity are possible and much more desirable. No attraction should be designed without some knowledge of what the users will have been exposed to previously.

Order and Relativity

The basic design principle of order, as set forth many years ago by Henry Vincent Hubbard and Theodora Kimball (1929), still applies:

> Our pleasure in the composition of a landscape depends on our appreciation of the ordered relations which exist among its parts. . . . The separate objects in the composition must be either harmoniously related to one another by repetition or sequence or balance.

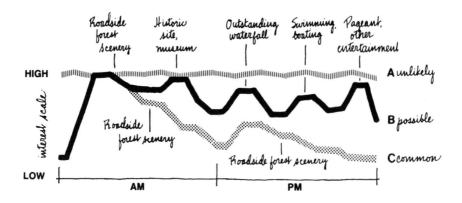

Figure 9-1. *Tourist attraction sequence. A diagram illustrating the typical tourist reaction to three levels of exposure and satiety: A represents the unlikely reaction of high interest throughout the day; B indicates a preferred sequence of peaks interspersed with lows; C is typical of satiation from too much of the same stimulus.*

A more contemporary source, landscape architect, J. O. Simonds (1985), has noted: "As human activity in an area increases, the landscape becomes more and more organized; agreeably if the organization is one of fitting relationships, disagreeably if the relationships are chaotic or illogical."

Order is not regimentation. Order is the arrangement of elements in ways that can be readily comprehended. Order grows from the logic of human use of the environment, removes the encumbrances of confusion, and assists the visitor in his physical and mental accomplishment of each day's travel.

For generations the free enterprise sector in market economy countries has cherished the right of independence and exclusiveness in its decision making. For many good reasons, much decision making occurs behind closed doors, and action is taken quite independently of other businesses and organizations—certainly separately from governments. As a process of business, this practice has ample defense. Some of its effects, however, place some limitations even on business itself.

Today, the hotel, motel, and motor-inn businesses, for example, do not really consider their function of overnight accommodations as related to travel objectives. Hence, they collaborate very little on matters of mutual concern, especially those regarding the surrounding environment. They are usually not associated, formally or informally, with those who establish attractions, even though attractions are most often the reason for their existence. Little linkage exists between lodging and transportation interests, in spite of their obviously interdependent functions. Each operates as an independent cell, and their managers hope that the other cells fit into place to form effective organs. Sometimes they do; often they do not.

This discussion gives rise to the principle of relativity in environmental design for tourism. No decision maker—whether in government, business, or a nonprofit organization—can establish any enterprise in this field today in isolation. All businesses, parks, attractions, and events that cater to travelers are to a greater or lesser degree related to one another and, especially, to the environment. Just as soon as a new element is established, the economics of supply and demand are altered, and so is the relationship among all elements of the environment.

As has been noted, this principle is not exclusive to business. Highway decision makers tend to ignore the effects of their decisions on the establishments that depend on highway transportation. Owners of huge reservoirs deny responsibility for what happens around their flowage property lines. Government park and recreational agencies prefer not to assume responsibility for development beyond the boundaries of state or national lands. By thus upholding a technical definition of jurisdictional responsibility, they severely restrict the opportunity for interrelated design. All these elements are integrated by society, and especially travelers, but this integration has yet to be reflected in policy.

Integration of attractions, services, transportation, information, and promotion is not easy, but it is necessary. In no other way can the ills and obstacles confronting overall quality tourism be eliminated. Achieving this end is one of the most important roles of designers and developers. More and more, they can use elements of design to integrate various functions into a mosaic. This unity cannot be created through isolated cellular design or decision making. Because of the principle of relativity, all stakeholders must cooperate.

Reuse

Probably no other form of economic activity makes a heavier demand on land than tourism. Markets are distributed to products, not the other way around. As a consequence, daily market shifts make it extremely difficult for the design of tourist places to remain static. If such places are to survive, the solution may lie as much in reuse of present sites and buildings as in new development.

To anticipate the capricious nature of markets, the designer must think broadly. Clients (owners and developers) may have short-range objectives. Can designers look far enough into the future to anticipate changes in demand? Perhaps. But, more likely, present development may have to be recycled from time to time to meet new needs.

The design needs of service businesses are different from those of attractions. Travelers seek higher and higher standards in accommodations, food services, and transportation. They want greater comfort, convenience, and speed, and better amenities. Advances in technology tend to render older facilities obsolete. However, historic reuse is proving that, for some attractions, old is better than new; designers and planners must understand that new hotels, restaurants, and service stations may be inappropriate in or near historic attraction clusters.

Designers in European countries have generally followed this design principle more strictly than those in the United States. Working with less land and greater restriction on sprawl, European designers and planners have had to use sites over and over again. In major cities, older residences and offices have often been converted to bed-and-breakfast accommodations, providing an appropriate location and design linkage with historic attractions.

Historic reuse is good economics. No community can afford to restore all old structures as museums. Through facade protection and other reuse mechanisms, contemporary uses of interiors can be developed while the older character of exterior design is retained. Besides, as new site and building costs continue to rise, older structures, even those needing new electrical, plumbing, and heating systems, often become economically viable by comparison.

Wholeness of Human Use

An overriding principle of tourism development today must be a greater sensitivity to the feelings and attitudes of travelers as entire persons. Those on pleasure or business trips are transplanted persons; they bring with them all the intellectual, emotional, and temperamental qualities they possess at home. They are not just campers, conservationists, fun-lovers, or sportspersons. They are individuals who, at a particular site at a particular time during travel, may wish to see or do a certain thing.

Mass affluence, mass travel, and mass recreation, with overemphasis on numbers, suggest the impersonal, the statistical, and the fleeting. Yet today, for the first time in history, millions of people are able to learn history, geography, politics, and sensitivity to other cultures not merely from books, radio, the press, computer communications, and television, but also from their own travel experiences. Whereas some of these experiences are frivolous and strictly for momentary personal gratifica-

tion, many are serious, memorable, enriching and of the land.

Critics of travel emphasize that tourists really absorb little today and, because of promotion, see what they are told to see. This criticism points to the great challenges designers and decision makers face: to lead people into strong and subtle environments and to provide them with experiences that are meaningful, pleasurable, and fruitful. Hence a major moral and ethical responsibility for the quality of the total environment and the wholeness of human use is becoming as important in the design of destinations for travelers as in the design of public places for local residents.

The thesis of this book is that design and development are driven as much by markets as by land resources. Therefore designers need to understand how people in the many market segments behave as much as they need to understand the physical materials of the environment. Attractions, services, transportation systems, visitor centers, and destination zones require design and development in congruence with all the activities of visitors that are important to their travel experience.

Innovation and Creativity

The qualities of innovation and creativity seem to override all others in importance. Because the lessons drawn from research are based only on past experience, they can offer no more than suggestions for future development. Certainly, the hazards of pollution and the public's distaste for ugliness are perpetually relevant facts. But with respect to the kinds of tourism environments that will be desired and used by people of the future, valid insight depends more on what is envisioned and created than on statistical interpolations and projections. Facts do not create design. Observed Chimacoff (1982), "The point is that there is neither control nor predictability about the ways in which inspiration strikes or the ways in which different people's minds assimilate, organize, and synthesize things."

Here lies the real challenge for designers. By intuition, experience, training, and clues from decision makers and research, they have the responsibility and the opportunity for solving design problems imaginatively. In their minds designers must see both the crowds and the solitary traveler, must see the processes by which travelers pass through and respond to their environments.

A word about practicality: Frequently the layman and the decision maker consider the designer's job impractical. Perhaps the business sector is most voluble in this criticism. Such an attitude stems from a failure to recognize the necessity for completeness and wholeness in any establishment that is to be used by people. This criticism implies that as long as the structure does not blow down, and as long as there is enough asphalt to park the car, all is complete and sufficient. If the structure were designed for robots devoid of any sensitivity to their environment, those features would be enough.

But the users are people—people with accumulated experiences, attitudes, emotions, and tastes. They like more than raw practicality. The evidence is in what they choose: They choose things that please their aesthetic and cultural senses while fulfilling a mechanical or physical function. The secret is to search out the ways and means of giving the public a choice, and here lies the golden opportunity for the business innovator and the designer. More of the same, or a duplicate of what in the past may have been successful, may be the very thing the business and public sectors should avoid in favor of the fresh, the sparkling, and the new. Clients should be made to understand that a designer's ability to visualize the completion of a project may go "beyond what tradition knows, likes, and feels unquestioningly at home with." (Zucker 1983).

The next chapter includes several examples that illustrate many of the observations, principles, and recommendations advocated in preceding chapters. Although each one is presented only briefly, the reader should examine each one carefully for the lessons it may offer.

The major issue before tourism development today is saturation—concern over the limits. It is essential that all planners and developers constantly monitor tourism to prevent degradation of resource quality and the visitor experience. Already, many popular destinations, such as Venice, face this critical issue. (Photo of Island of St. George courtesy Italian Government Travel Office)

CHAPTER 10

Gallery of Examples

ncreasingly, community and destination tourism planning and development examples are arising to demonstrate desirable approaches. Following are some cases that illustrate elements of worthwhile processes, techniques, and designs, even though they are directed toward various tourism purposes.

DEVELOPMENT

This group of cases illustrates applied approaches to tourism development. Each section presents the purpose and method by which tourism development has been approached and carried out in different locations. Although these cases vary in many respects, the reader should be able to observe some common threads that may be applicable to other situations.

Egypt

The case of the west coast of the Red Sea and South Sinai of Egypt demonstrates a new approach to planning, design, and development (Hamed 1991). This area, as the crossroads of Asia, Africa, and Europe, has been a favorite travel destination for centuries. Existing tourism, although economically productive, has shown a dark side. Resentment among some local populations, natural resource degradation, and social impacts have stimulated the new approach. The first manifestation of needed change was a recognition of the many players in tourism development: the Egyptian Ministry of Tourism, international funding institutions, private-sector investors, related municipal agencies, members of the construction industry, and tour operators. The government of Egypt in 1989 sought assistance from the United Nations Development Program, the U.S. Agency for International Development, and the World Bank.

A preliminary study resulted in the creation of a new responsible group, the Tourism Development Unit (TDU). Landscape architect Safei El-Deen Hamed was engaged to provide answers to: "What would you do to ensure that the proposed investment in tourism ultimately provides an environmentally sound physical development plan for each of the priority areas selected for tourism?" An initial step in answering this question was the establishment of the Environmental Affairs Division (EAD) of the TDU. The main functions of this division are "to assess and monitor all projects and programs from the natural and cultural resources point of view." Three major layers of decisions are involved.

Strategic decisions encompass five basic elements:

1. A basic mission is to integrate tourism development into the long-range sustainability of resources. EAD's responsibility includes defining goals, reviewing projects, monitoring development, and making recommendations.
2. The identification of EAD's clients was essential and has resulted in naming tourism developers, conservation organizations, governmental agencies, and consultants as most critical.
3. Stating goals and objectives has involved observing existing works, screening past statements, and formulating new goals and objectives within the scope of EAD.
4. If this organization is to be effective, it must identify its programs and sources, survey existing data, develop environmental standards, inventory the available resources, establish regional advisory boards, supervise construction sites, review applications, provide education, and monitor developments.

A second layer of the TDU's decision making involves coordination. This function requires integration into the overall national development plan. Factors to be considered include national policies on tourism, national building codes, national environmental laws and all relevant land use plans.

A third layer of decisions to be made involves human resources needed, time spans required, and budgets for staff and operations.

This example illustrates the need for strong ties between designers and planners and a nation's various government entities for better tourism planning and development.

Mount Desert Island

A small island off the Maine coast, Mount Desert Island contains about 10,000 inhabitants scattered in four communities—Bar Harbor, Mount Desert, Southwest Harbor, and Tremont. In recent years many residents became concerned about growth, especially tourism growth resulting from increased numbers of visitors to a major attraction, Acadia National Park (Beard 1993). Additional segments of the local economy include boat building, fishing, and a scientific research center.

After contacting several agencies and organizations, local people concerned about growth issues obtained the cooperation of a specialist from the University of Maine Cooperative Extension Service. As a result of bringing together representatives of key organizations, in 1991 they formed a steering committee and adopted the title of Mount Desert Island Tomorrow (MDI) with the following purpose: "To help MDI citizens and communities manage cumulative, island-wide impacts of growth, identify and build consensus about the island's future, and cooperatively guide development so as to protect and improve environmental, economic, and social conditions."

Six key issues were identified, all related to tourism: sewage and solid waste disposal; quality and supply of drinking water; protection of open space and agricultural land; traditional access to the shore and uplands; housing, land costs, and the property tax structure; and traffic and transportation.

After many communitywide meetings, a sixteen-page report titled *MDI Tomorrow: A Look to the Future of Mount Desert Island* was prepared. Even though no new organization has evolved, a loose structure, led by the extension specialist and the steering committee, has made several notable accomplishments. Improved housing, new plans for guiding growth, a comprehensive plan for Bar Harbor, improved visitor services, and new collaboration among the several public and private organizations have resulted from the group's efforts. For the first time, service clubs, elected officials, business leaders, planning committees, and chambers of commerce have begun to work together.

Conclusions reached after a few years are as follows:

- Community attitudes and values are the basis for the work of the community.
- There is a resident capacity to identify and solve problems and meet the needs of community members.
- The organizational structure of the community encourages participation of all its members.

- The community has the means to build, maintain, and renew leadership.
- Community members have the means to build consensus and articulate a shared vision about a preferred future.

This case demonstrates how local residents, without input from outside investors or government bureaucracies, can resolve their own issues and guide their own destiny, provided that a catalyst—in this case the extension specialist—is present to supply the needed unbiased leadership.

Caribbean Island Communities

Darrow (1995) has proposed a process of six phases for planning tourism at the level of the Caribbean island.

Phase 1 involves the selection of partners, such as the United States, for coaching, education, and financial assistance. Despite such partnerships, emphasis is to be placed on maintaining local control. All partners must understand the relationship between market preferences and the local resources for development.

Phase 2 is the establishment of roles for all partners. This step is also one of establishing trust, commitment, and teamwork.

Phase 3 is a refinement of the previous step and includes articulating a "shared vision," the major purposes of development. This phase also places great emphasis on integration with local traditions and activities.

Phase 4 is the planning stage that includes establishing goals and specific objectives within each goal. For each objective, the organizations and agencies to take action are identified. Establishment of an implementation schedule is critical.

Phase 5 includes continued partnerships for evaluating each action taken, not merely within each project but throughout the community and surrounding area. The interests of all parties are safeguarded at this step by means of communication and cooperation.

Phase 6, the final phase, provides for monitoring progress. Such monitoring is to include tourists' reactions as well as the experiences of the action parties. Modifications or complete changes in direction should be the result of this continuing phase.

Wolfville, Nova Scotia

Wolfville (pop.3500) is located on the north shore of Nova Scotia, approximately ninety kilometers from Halifax. Settled in 1760, it is now the home of Acadia University and has many points of interest to tourists on the Evangeline Trail. A volunteer group of interested local parties conceived the project *Destination Wolfville* (Town of Wolfville 1994). Adapting existing tourism development guidelines (Murphy 1985, Gunn 1979) to local conditions, the project members created the model illustrated in Figure 10-1. Further explanation is provided in the following process steps, developed by the citizens assisted by Acadia University:

1. *Community involvement.* Community action: town council support, assembly committee meetings, public meetings, definition of Action Committee's tasks.
2. *Data and information gathering.* Situation analysis: environmental analysis, resource analysis, tourist analysis. Community inputs and involvement: focus groups, interviews, workshops.
3. *Evaluation of data and information.* Community goal formulation: general planning philosophy, SWOT analysis, community planning goals, community themes.
Community strategy foundation: strategic objectives, community products and services analysis, tourist market identification, service market strategy. Priorities for tourism: program development, physical development.

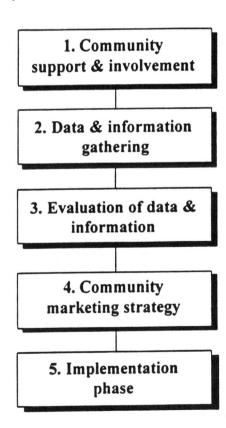

Figure 10-1. *Wolfville tourism planning model. This diagram illustrates the process of planning and programming for tourism development in Wolfville, Nova Scotia. Source: Destination Wolfville 1994.*

4. *Community marketing strategy.* Community marketing strategy: marketing mix, products and services, pricing strategy, channels of distribution, promotional plan. Target marketing strategy: market identification, primary research data generation, segmentation strategy. Community positioning strategy: theme verification, image identification.
5. *Implementation.* Community organization design: implementation committee, composition and structure. Implementation policies: resources, sites, and programs, major existing sites and programs, additional resources. Management support systems: community information systems, community planning process, monitoring and evaluation systems.

In the past the tourism development had catered primarily to touring circuit travelers, such as individual parties or motorcoaches that made the town a stop on a tour. A major purpose of this new strategy was to develop a stronger concept of destination. A key feature of the process was to involve a great diversity of players at the beginning step and throughout the process.

A Pacific Northwest Village

Professional planners and landscape architects Lankford and Lankford (1995) established a framework for community tourism development and applied it to a Pacific Northwest village. Their approach stressed the need for strong public input from the very beginning of any tourism development. These authors have described their four-phase approach:

1. *Review of existing plans and studies:* analysis of present community factors based on surveys, public meetings, interviews, and secondary sources. Topics included: land use, physical and natural characteristics, recreation resources, visibility, signage, and the sociodemographics of residents.
2. *Market analysis:* review of current and potential demand, including origins, visitor characteristics, seasonality, adequacy of accommodations, and potential future markets.
3. *Transportation strategies:* circulation, access, parking needs; pedestrian needs and issues; and visibility and signage.
4. *Development options:* Four main opportunities for development, community design concepts, and strategies for future development and financing.

Arizona

In 1993 the Governor's Rural Development Conference in Arizona asked delegates from rural communities what was required for enhanced rural tourism development. As a result, the Arizona Council for Enhancing Recreation and

Tourism (ACERT) was formed (Andereck 1995). This was the first time that such a council had been formed in the United States. Its purpose was "to improve recreation and tourism in Arizona and to foster increased cooperation and coordination among federal and state agencies, Indian nations, and private industry" (Andereck 1995, 9). Cooperating agencies and organizations contributed staff time and financial resources for administration of the program, as illustrated in Figure 10-2.

It was agreed that plans for tourism development in the state need to be:

1. *Comprehensive.* Plans must cover all aspects of tourism development and management.
2. *Sustainable.* Development projects must protect community values for the long term.
3. *Low-cost.* All development must be supported by the community and participating groups for the long term.
4. *Customer-oriented.* Customer service must prevail and be founded on state-of-the-art processes.
5. *Fiscally positive.* Community revenues from development must be encouraged.
6. *Technically sound.* Plans must provide linkage with governmental and university programs.

The operation of this program at the local level depends on the action of several teams. A *local community action team* of local representatives schedules participation of other teams and evaluates proposals. A *reconnaissance team* is made up of agency specialists who study the local potential and make recommendations to a *resource team*. This team analyzes strengths, weaknesses, and potential for tourism. A *facilitation team* of agency specialists guides the preparation and implementation of action plans.

The process utilizes several steps:

1. *The community action team* is required to identify its competence to work on the project—a *pre-acceptance diagnostic.*
2. Based on its evaluation, the ACERT approves the selection of the community for support.
3. After reviewing the *pre-acceptance diagnostic,* the *reconnaissance team* lays the groundwork for a community visit by the *resource team.*
4. Several teams make an assessment of potential during a two- or three-day visit.
5. If the values of the community are compatible with tourism expansion, goals for development are prepared.
6. Next is the creation of an *action plan.*
7. Specific strategies are identified to guide development.
8. ACERT specialists provide assistance in details of individual projects.
9. The final step is *evaluation and accreditation,* to monitor change as it takes place.

An application of this process was begun in 1995 in the Globe-Miami contiguous communities east of Phoenix. After the *reconnaissance team* met with local

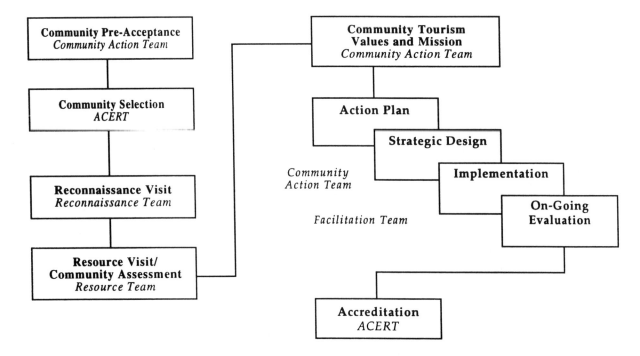

Figure 10-2. *Arizona rural tourism development program. This diagram illustrates the steps for evaluation and recommendations.* Source: Andereck 1995.

Study of the area around Globe-Miami, Arizona, revealed many opportunities for new attraction development, such as these Indian ruins nearby. (Photo courtesy Arizona Department of Commerce)

tourism interests, recommendations on areas of expertise were made for selecting *resource team* members. Following a comprehensive visit by the *resource team*, assessment was made for the following (ACERT 1995):

1. *Organizational structure.* Factors rated on a four-point scale included: an active tourism organization, the ability to raise funds, trained staff, collaboration with other communities, participation in regional efforts, public–private partnerships, a monitoring system, and tourism plans and programs.
2. *Tourism resources.* Factors evaluated included: attraction inventory, resource attraction capacity, coordinated programs, public land resources, and profitable festivals and events.
3. *Tourism businesses.* These were rated for the following factors: availability, quality, and variety; new business potential; public visitor services; customer service orientation; operating hours; and foreign exchange and automated teller machine capabilities.
4. *Marketing program.* Again on a four-point scale, from superior to nonexistent, the following were evaluated: promotional mix, documented traveler characteristics, evaluations of visitor satisfaction, marketing plan, quality of marketing materials, responsiveness to requests, and internal marketing program.
5. *Community involvement.* Factors assessed included: quality of the citizen involvement process, positive community values and attitudes, and responsiveness to tourism professionals.
6. *Community leadership.* Key issues evaluated included: community volunteerism, entrepreneurial leadership, and political commitment.
7. *Community design.* Encompassed in this evaluation were: uniqueness and appeal; entry; signage quality;

parking, sidewalks, and parks; and functional relationships and architectural sensibility.
8. *Sustainability.* Factors assessed included: public revenue, business profits, human resources, physical resources, and capacity for change.

This application is in the process of moving into the next phases, which will address the community's several needs as determined by the assessment of the *resource team*.

An Ecotourism Company

Ultimate Adventures, an ecotourism project in northwestern Canada, is demonstrating how a private commercial enterprise can be successful at the same time it protects the environment and operates in a sound ethical manner (Wight 1995).

The product includes guided hiking, canoeing, mountain biking, climbing, and caving. But perhaps most important to the product is the way in which these activities are managed. For example, the enterprise's environmental policy covers such details as: preparation for tours, activities at camp, campfire activities, cat holes (latrines), washing activities, and litter and garbage disposal. The company mission statement follows.

> To provide our clients with quality programs and outdoor adventure experiences that may enhance our appreciation of the environment and to promote environmental awareness in a way that we take only pictures and leave nothing but footprints.

The proprietor, Andrew MacKenzie, offers the following recommendations for development of projects such as this one:

1. *Search for information.* Know your objective, study examples elsewhere, and turn problems into opportunities.
2. *Transform information into new ideas.* Adapt findings to the case at hand, consider linkages with community service, and eliminate project ideas that are not feasible.
3. *Evaluate ideas and determine action.* Factors to be examined include: foundations worth building upon, potential drawbacks, timelines, and probabilities for success.
4. *Carry ideas into action.* These steps are essential: Prepare a precise plan, be driven to succeed, understand the effort and costs needed, establish physical development and programs, understand the product, be determined, and follow through.

Many policies and practices were established in the course of this project, such as for: equipment needed, proper use of trails and sites, avoidance of fragile resources, proper use of camping areas, waste and disposal treatment, and proper washing methods. The owner-manager has taken care to apply these policies; he attributes his success to these principles: cooperative networking with relevant agencies and organizations, working closely with the local community, establishing partnerships such as with airlines and travel agents, researching needed information, and learning from mistakes.

This case demonstrates many principles that are applicable elsewhere.

A Scenic Byway

A demonstration project sponsored in part by the U.S. Department of Agriculture Natural Resources Conservation Service (NRCS) is the Loess Hills Scenic Byway in southwest Iowa (U.S. Department of Agriculture and National Endowment for the Arts 1995). The service works with local people through resource conservation and development (RC&D) areas. The Loess Hills project was led by the Golden Hills RC&D, a group of eight counties with a total area of 3 million acres and a total population of 190,167 residents.

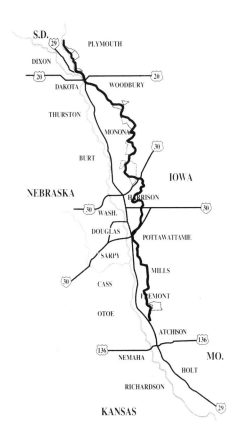

Map of the 221-mile Loess Hills scenic byway with twelve excursion loops. Future sustainability is governed by a corridor management plan for conservation and development of resources along the route. (Plan courtesy USDA Natural Resources Conservation Service)

Because of drops in the farm economy in the 1980s, local people began looking toward other sources of income. Over a period of four years, the project to identify, analyze, and designate the Loess Hills Scenic Byway came into being. Following are the key steps of this grass-roots project.

- Landscape architects Ronald W. Tuttle and Mimi W. Askew of the NRCS began local leadership in 1989 and created sketches for a potential scenic route.
- The first segment was completed in 1990 with strong sponsorship from the Western Iowa Tourism Region. Problems encountered included poor attention to circulation, little cooperation with the Iowa Department of Transportation (IDOT), and a poor brochure.
- Through many local meetings led by Askew, the route was expanded.
- For guidance on route selection, NRCS developed an official procedure for assessing the quality of countryside landscapes. This process defines three basic qualities: *character:* a measure of how well the landscape and its components work or belong together; *structure:* a measure of the distance that one can see in a view at any one time; and *information:* a measure of the landscape's ability to engage the viewer.
- County road engineers and IDOT evaluated safety issues related to selected routes.
- Further analysis included study of the suitability for support from tourism.
- Final route selection was made, primarily as a result of public involvement throughout the process.
- The Loess Hills Alliance, an organization representing all counties involved, was given the responsibility for resource protection throughout the corridor.

Local people are beginning to see the results in their communities as more and more visitors are taking the Loess Hills Scenic Byway. This project is stimulating local interest in resource protection and recognition of the value of a balanced economy. It also demonstrates that the combination of technical expertise and local citizens' involvement is a viable means to accomplish desired objectives.

An Ecomuseum

Kalyna Country, an area of 15,000 square kilometers east of Edmonton, Alberta, has established a new ecomuseum and destination complex that protects and interprets local natural history, cultural history, and artifacts of early Ukrainian settlement of the area (Tracy 1994). Six major themes were chosen by the museum's board of directors: geology, physiography, and glacial history; the natural environment, past and present, as reflected in both the fossil record and in contemporary flora and fauna; aboriginal history, culture, and modern

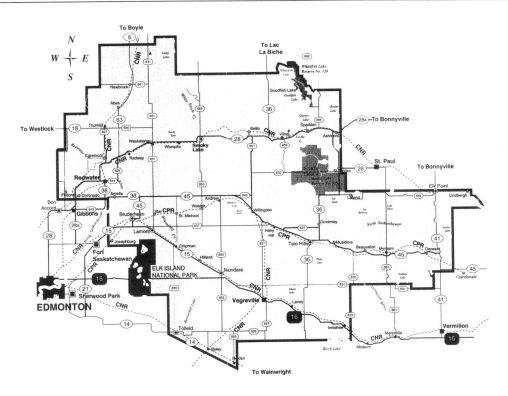

The land area included in the cultural tourism attraction complex, the Kalyna Country Ecomuseum, located east of Edmonton, Alberta. (Map courtesy Kalyna Country Ecomuseum)

native life; the period of European expansion and the fur trade; early agricultural settlement and the pioneering experience; and the modern contemporary life of local Ukrainian communities.

This project was initiated by Alberta Community Development (a provincial agency) and the Canadian Institute of Ukrainian Studies at the University of Alberta. The board of directors includes representatives of Cree, Romanian, English, and Ukrainian ethnic groups in the area. Restoration of historic buildings and landscapes is taking place in this living history museum, which is boosting the local economy through tourism and yet protecting natural resources and the rich heritage of the region.

Key features of this areawide tourism development include: twenty-three different local museums, a recreated Ukrainian Cultural Heritage Village, Elk Island National Park, Victoria Settlement, Beaverhill Lake (for bird watching), the Fort George/Buckingham House interpretive center, the homes of modern-day descendants of aboriginal inhabitants, more than one hundred historic churches, and many festivals depicting the several cultural facets of the region (Kalyna County Ecomuseum n.d.).

An Ecolodge

An example of a lodge that was planned and managed in an environmentally sensitive manner is Purcell Lodge, located in the Canadian Rockies at Golden, British Columbia (Alberta Tourism 1994). The purpose of its location

was to provide for a specialized travel market segment interested in hiking and wildlife viewing in the summer and backcountry skiing in winter. The lodge is accessible only by helicopter and accommodates twenty-four people.

The setting is wilderness; the accommodations are luxurious simplicity. Fine dining, deluxe rooms, a social room with a fireplace, a quiet reading area, hand-crafted furniture, and a sauna for weary hikers and skiers are

A major feature of the Kalyna Country Ecomuseum is the redevelopment of a Ukrainian village, including this typical Ukrainian church.

features of the lodge. It offers visitors a great diversity of year-round activities in the surrounding area. In accord with the wildland theme, there is no television, radio, telephone, fax machine, or cellular phone network. Guides are available for nature interpretation as well as outdoor sports instruction. The lodge is located at an elevation of 2,195 meters in the foothills of the Rockies.

Many new technologies have been used at the lodge in efforts to respect environmental constraints and to increase efficiency. For example, its total electrical power consumption is equivalent to that of two personal hair dryers. This low consumption is made possible by a 12-kilowatt hydro-turbine generator for forced air heat, showers, and electrical service. An electronic load control governor distributes electricity only at times of need.

For waste management, low-flush toilets discharge into a self-contained aerobic system on site. The steps include a septic tank, an aeration tank, a chlorinator, a holding pond, and a leach field. Effluent eventually seeps into the groundwater supply. Grease is trapped and disposed off-site.

The systems have the advantage of annual operation at relatively low cost. However, the initial investment for the lodge was relatively high.

A Waterfront Plan

An example of a concept for an urban waterfront tourism development is outlined in the *Fisherman's Cove Waterfront Development Plan* (Shearwater Development Corporation 1995) for Halifax, Nova Scotia. The proposal is presented in six parts: research, public participation, development issues and opportunities, waterfront plan concepts, waterfront development plan, and implementation. The proposed location is adjacent to the small town of Eastern Passage, across the harbor from Halifax.

Research for the proposal included identification of natural and cultural resources. The nearby islands of Lawlor's Island, McNab's Island, and Devil's Island are scheduled to become a provincial park. Lawlor's Island has become a major rookery for the great blue heron. Waterfront vistas are important, and historic ruins at the site include a nineteenth-century Quarantine Hospital. The serene natural setting is in direct contrast to the industries and urban development of Halifax and Dartmouth. Some improvements in sewage disposal will be needed. Access is convenient and direct; the area is serviced with electricity; water is available; and commercial fishing can continue, but on a limited basis.

For public involvement, the planning team met directly with key residents for one-on-one interviews, gave a public presentation of issues and opportunities, and gained responses to a draft development plan. An *Eastern Passage Tourism Development Opportunities Report* was prepared, which requested opinions of residents of the area. Resident recommendations included to build a cultural interpretive center, to resolve current conflicts over parking and wharf access, to repair fishing sheds, to develop better recreational boat access, to improve identification and signage, and to create better pedestrian linkages and other enhancements. Potential benefits cited included retention of the area's fishing village character, improvement in visitor interpretation and demonstration, construction of new boat launching facilities, construction of marine infill, and improvement in entrance design.

The concept plan incorporated many citizen recommendations as well as professional design input. Integration of new tourism and local recreation with existing buildings, wharves, and community activities dominated the planning approach. Projects included a new interpretive and demonstration center, improved pedestrian circulation, and restoration of traditional and indigenous features as attractions.

Purcell Lodge, Golden, British Columbia, is an ecolodge development that offers deluxe facilities in a wildland setting. The design utilizes new environmentally sensitive technology. (Photo courtesy Purcell Lodge)

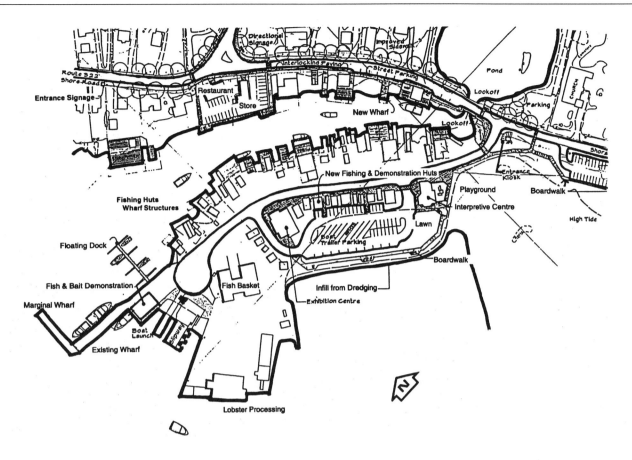

An old fishing harbor is to be redeveloped for local and tourist visitors according to the Eastern Passage Waterfront Plan for Halifax, Nova Scotia. Both natural and cultural resources are to be interpreted. (Plan courtesy Fisherman's Cove, Halifax)

The waterfront development plan encompassed nine basic attractions: fishing village, interpretation and demonstration centers, fishing village support, pedestrian circulation, Quigley's Corner, McCormick's Pond, general landscape development, Lawler's Island, and a marsh boardwalk exhibit.

Dewees Island

An environmentally sensitive vacation home development, Dewees Island (1995), South Carolina, is proving that care in planning is good business and marketable. This 1,200-acre island has 2.5 miles of beach, a wildlife impoundment area, lakes, abundant bird life, and a nature preserve. Only 35 percent of the area is platted, for a permanent maximum of 150 homes. Rigid construction rules control land use, architecture, and all engineering systems. Access roads are restricted to electric vehicles, and there is no golf course. The island is accessible by a nine-minute ferry ride from the Isle of Palms, adjacent to the South Carolina coast near Charleston.

Rather than dampen market demand, the strict environmental controls and building restrictions are enticing buyers. Homes may not disturb more than 7,500 surface square feet, less than 9 percent of most lots. All home-

owners must abide by the guidelines established by an architectural resource board, including vernacular architectural themes. Passive heating and cooling techniques, water conservation fixtures, and waste composting and recycling make virtually no negative impacts on the environment.

Initiated in 1991, the project plans and management resulted from input from equity investors, original landowners, designers, environmentalists, and developers. The

Sketch of waterfront wharf and fish shed improvement, part of the redevelopment plan for the Eastern Passage area. (Sketch courtesy Fisherman's Cove, Halifax)

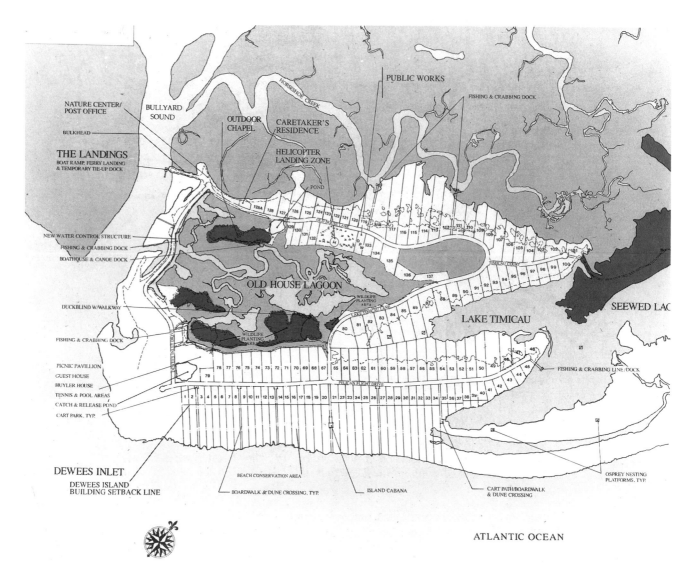

NATURE CENTER/
POST OFFICE   BULLYARD
                     SOUND

BULKHEAD

THE LANDINGS
BOAT RAMP, FERRY LANDING
& TEMPORARY TIE-UP DOCK

OUTDOOR
CHAPEL   CARETAKER'S
                RESIDENCE

HELICOPTER
LANDING ZONE

HORSESHOE CREEK

PUBLIC WORKS

FISHING & CRABBING DOCK

POND

NEW WATER CONTROL STRUCTURE
FISHING & CRABBING DOCK
BOATHOUSE & CANOE DOCK

OLD HOUSE LAGOON

DUCKBLIND W/WALKWAY

WILDLIFE
PLANTING
AREA

SEEWED LAC

FISHING & CRABBING DOCK

WILDLIFE
PLANTING
AREA

LAKE TIMICAU

PICNIC PAVILLION
GUEST HOUSE
HUYLER HOUSE
TENNIS & POOL AREAS
CATCH & RELEASE POND
CART PARK, TYP.

PELICAN FLIGHT DRIVE

FISHING & CRABBING LINE/DOCK

DEWEES INLET
DEWEES ISLAND
BUILDING SETBACK LINE

BEACH CONSERVATION AREA

BOARDWALK & DUNE CROSSING, TYP.

ISLAND CABANA

CART PATH/BOARDWALK
& DUNE CROSSING

OSPREY NESTING
PLATFORMS, TYP.

ATLANTIC OCEAN

Layout plan for Dewees Island vacation home project. Only 35 percent of the area is scheduled for building development; the remainder is to be preserved in its natural state. (Plan courtesy Dewees Island, South Carolina)

island demonstrates the balance between demand and supply of quality resource development.

A Shopping Village

An old mill town with a population of three thousand in Ontario, Elora has been successfully converted to a tourist shopping village (Getz 1993). The old mill has been made into an up-market inn and restaurant, and old structures now are filled with antique shops, art galleries, craft boutiques, and gift stores. Several older homes have been converted to bed-and-breakfast establishments. In the village an old church has been converted to ceramic and pottery production. Nearby is a park with campsites provided by the Grand River Conservation Authority. Several annual events have become popular, such as a music festival and an antique show and sale.

Early studies of visitors to the area cited scenery (the Elora Gorge of the Grand River), country atmosphere, peace and quiet, and the historical nature of the village—followed by shopping—as the most important attractions. A recent survey has shown that some segments of the travel market believe that the prime attractions remain but that better planning is needed to avoid congestion and an atmosphere that is too commercial.

This and similar studies of shopping villages have identified key development issues as: avoiding displacement of local shops by those that are tourist-oriented, retaining historic architectural quality as reuse takes place, maintaining local control, linking infrastructure to attraction planning, and capacity considerations.

A Rainforest Tourism Project

To provide nature tourists access to its unspoiled rainforest plant and animal life, the Canopy Walkway has

An example of careful siting of vacation homes on Dewees Island. Homes may not disturb more than 7,500 square feet of land area. Strict architectural guidelines ensure sustainable development. (Photo courtesy Dewees Island, South Carolina)

The Canopy Walk, Kakum National Park, Ghana, is one of a few such rainforest installations in the world. This treetop walkway with observation platforms gives tourists a rare view of jungle life. (Photo courtesy Conservation International and Ghana Tourist Board)

The Elora Falls of the Grand River swirl around a remaining rock centerpiece, a popularly photographed scene. The grist mill of the 1830s on the left is now restored and contains a bed-and-breakfast accommodation.

Representative species of wildlife that may be observed in Kakum National Park, Ghana. (Sketch courtesy Stephen Nash, Conservation International)

been established in Kakum National Park, Ghana. This is one of several such high-elevation walks in various parts of the world—Malaysia, China, Peru, and Australia (Conservation International n.d.).

This unusual walkway is clamped to eight large trees and stretches 333 meters between entrance and exit trees and huts. Included are six viewing platforms, the highest of which is 26.5 meters off the ground. A special design allows suspension without damage to the trees. To avoid excessive stress on the system, quotas are placed on the number of visitors allowed at any one time. The site was chosen by designer Illar Muul based on soils, slope, and especially quantity of hardwood trees sufficient for the project.

The rainforest in this region has a layered appearance, with the canopy emerging above all other vegetation. Some trees are 30 to 45 meters in height, shielding a variety of growth at lower levels.

Because of the rare high vista, visitors are afforded a spectacular view of plants and animals. A great diversity of trees, vines, mosses, ferns, and orchids can be observed. Song birds, screeching insects, and monkeys with their frequent strange calls can be seen and heard. Sound recordings of more than 116 bird species have been collected, from which a final recording will be made available. This project is an important part of the overall attraction complex of Kakum National Park.

DESIGNS

The following set of examples contains more site-specific tourism projects than the preceding section. They represent exemplary designs executed by architects, landscape architects, and others. These projects demonstrate aesthetically satisfying as well as functional solutions to tourism site challenges. Implied in these examples is the joining of travel market demand, owner reward, and environmental sensitivity.

Point State Park, Pittsburgh, Pennsylvania

GWSM, Landscape Architects

A generous open space in the heart of an industrial city, Point State Park provides dramatic vistas as well as repose. The design is large in scale, yet marries express-way travel to pedestrian traffic and offers memorable views of the confluence of the historic Monongahela and Allegheny rivers. The well-placed fountain, in scale with the entire setting, provides an accent in motion.

Broad river valley and fountain accent at Point State Park (below). (Photo courtesy GWSM, Inc.)

Cantilevered overlook at Point State Park offers access to dramatic views. (Photo courtesy GWSM, Inc.)

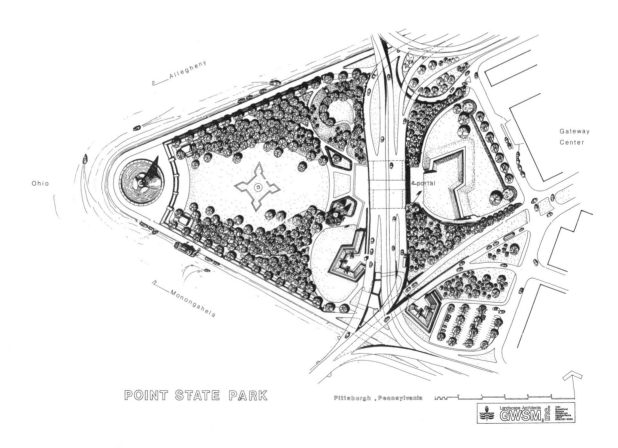

Plan of Point State Park showing design compatibility between expressway and pedestrian use. (Drawing courtesy GWSM, Inc.)

South Dunes Interpretive Walk and Beach, Jekyll Island, Georgia

Robinson Fisher Associates, Landscape Architects

Once a millionaire's playground, this island is now managed by Georgia's Jekyll Island Authority. Following a new land use policy for the entire island, which had suffered badly from erosion, the designer's objectives for the beach area were to restore the stability of the dunes, recreate the freshwater slough, and re-educate the public about the nature and value of the island's environment. Since redesign, the dunes are stabilizing, people are enjoying greater access to the beach, and the natural beauty and ecology of the area are being restored.

General plan showing relationship between restored dune area (below), redeveloped slough, and interior of South Dunes Park. (Drawing courtesy Robinson Fisher Associates)

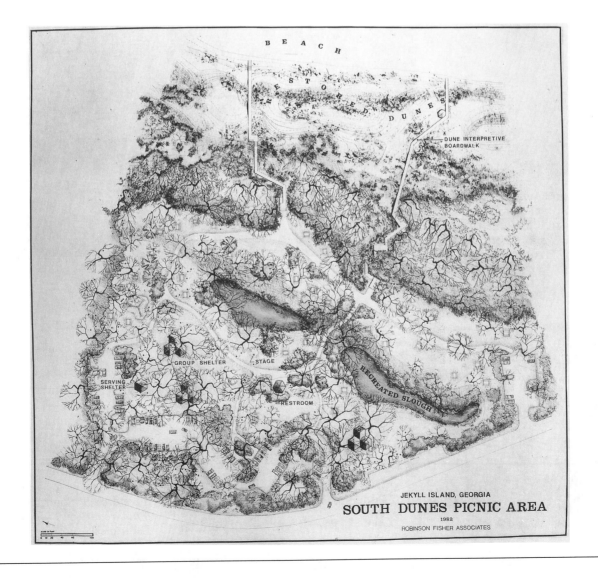

Group shelter area on less sensitive resource portion of newly designed South Dunes Park. (Photo courtesy Robinson Fisher Associates)

Design of boardwalk solves dune erosion problem and provides for an interpretive adventure at South Dunes Park. (Photo courtesy Robinson Fisher Associates)

Overall view of beach and dune area of South Dunes Park. (Photo courtesy Robinson Fisher Associates)

Miner's Castle Overlook, Munising, Michigan

Johnson Johnson & Roy/inc., Landscape Architects

This redesign of a heavily used resource area accommodates large numbers of visitors while arresting the erosion that has occurred in the past. Miner's Castle Overlook is only one part of an attraction complex, the Pictured Rocks National Lakeshore, owned by the U.S. National Park Service. Several new overlooks, a comfort station, new trails, a relocated entry drive, and facilities adapted to the handicapped are part of the overall design, which mitigates visitor impact to retain landscape quality.

Visitor trampling of sandstone at Miner's Castle had badly eroded this popular attraction. (Photo courtesy Michigan Tourist Council)

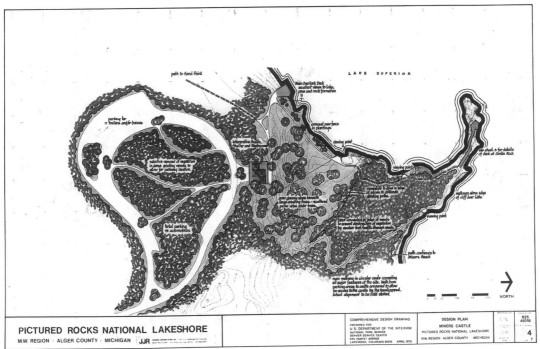

Overall plan of Miner's Castle illustrates efforts to reduce visitor impact and allow increased volume of visitors. (Drawing courtesy Johnson Johnson and Roy/inc.)

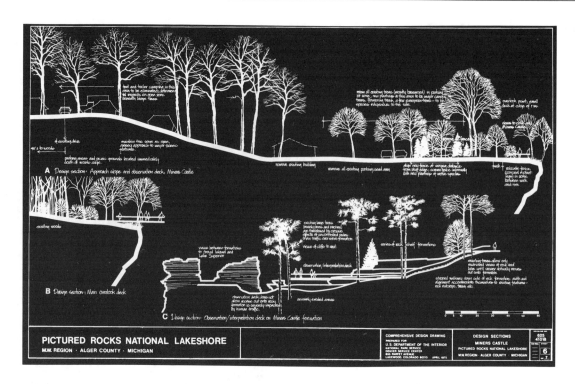

Studies of improved design of walks, overlooks, and relocation of camping at Miner's Castle. (Drawing courtesy Johnson Johnson & Roy/inc.)

This design solution at Miner's Castle has stopped further erosion, yet allows visitor access. (Photo courtesy Johnson, Johnson, & Roy/inc.)

Woodland Park Zoological Gardens, Seattle, Washington

Jones & Jones, Landscape Architects

Zoos are important attractions for visitors, especially when they are designed as well as this one. In Woodland Park, each exhibit, such as the African savanna exhibit and the gorilla exhibit, was situated for best fit between the needed bioclimatic zone and the existing vegetation and microclimatic conditions. Further design considerations were geared toward the animals' welfare first and the visitors' desires second. These priorities have assured more suitable habitats and enabled visitors to be immersed in the exhibit landscape. The designer's goal, executed with great skill, was to take people from their own backyards and "transport" them to exotic regions.

Visitors to Woodland Park feel as if they are original discoverers of exotic animals (above). (Photo courtesy Jones & Jones)

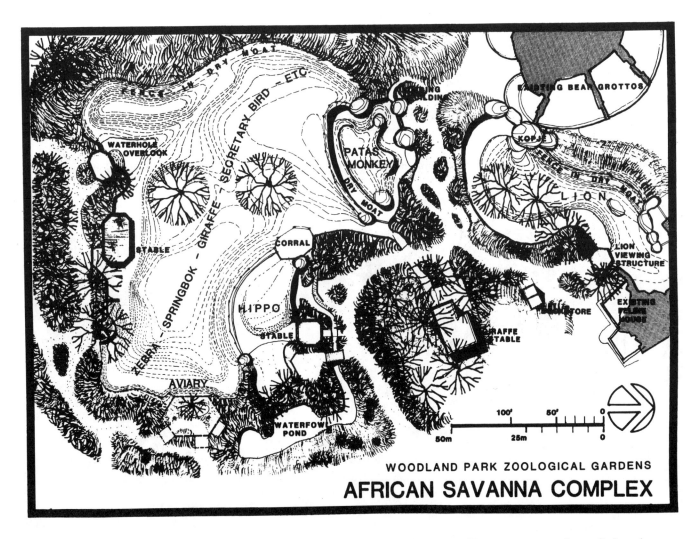

WOODLAND PARK ZOOLOGICAL GARDENS

AFRICAN SAVANNA COMPLEX

Plan of Woodland Park showing dominant animal habitat and ample public access. (Drawing courtesy Jones & Jones)

Landscape setting at access points sets the theme for exhibits at Woodland Park. (Photo courtesy Jones & Jones)

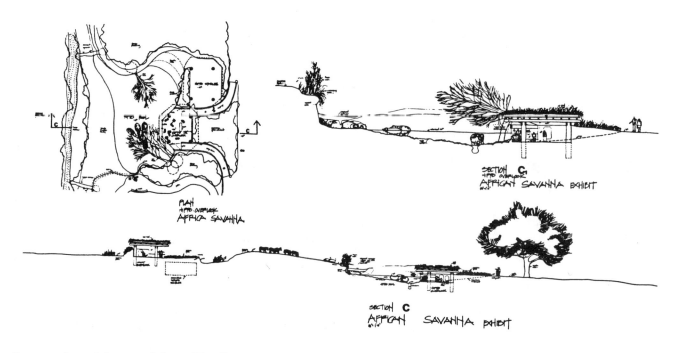

Sections through hippo exhibit at Woodland Park. (Drawing courtesy Jones & Jones)

Gas Works Park, Seattle, Washington

Richard Haag Associates, Inc., Landscape Architects

The designer's concept of retaining old gas works in a public park, originally opposed both by city officials and the public, has now proven to be the centerpiece of an award-winning design. The uncluttered views of the city and distant Cascade mountains have been retained. Original machinery, made safe for children to climb on, was painted in strong colors to stimulate their interest. An old boiler was fitted with a small stage and dance floor, demonstrating how imaginative design can reclaim abandoned industrial artifacts. The park is now used by tourists as well as residents. It also shows how effective a designer's catalytic role can be, bringing the several stakeholders together for a final design solution.

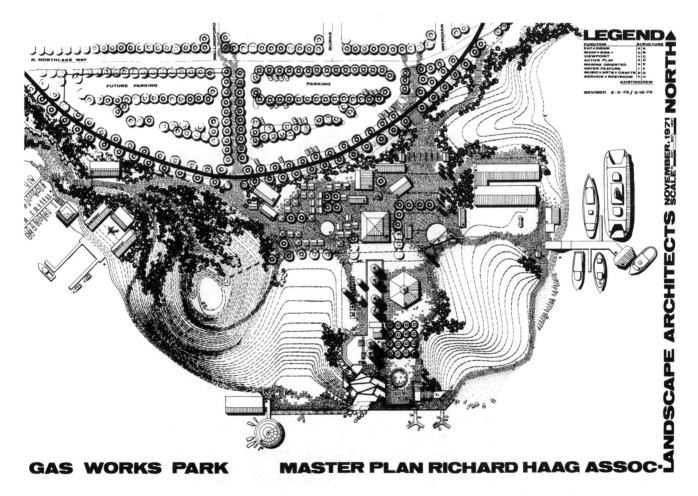

GAS WORKS PARK   MASTER PLAN RICHARD HAAG ASSOC.

Plan of Gas Works Park shows creative design of open park space and waterfront as well as reuse of gas works. (Drawing courtesy Richard Haag Associates)

Gas works structures and waterfront on Lake Union at Gas Works Park. (Photo courtesy Richard Haag Associates)

Gas Works Park is a fun place for children, as well as educational for all. (Photo courtesy Richard Haag Associates)

Jordan Pond House, Mt. Desert Island, Maine

Woo & Williams, Architects, Landscape Architects

Integrating public and private interests and blending site with building were the design challenges at Jordon Pond House. This gateway facility within Acadia National Park, which has a restaurant, gift shop, and viewing areas, captures the essence of the original Jordan Pond House, which burned down in 1979, and is enhanced by the natural beauty of its setting. The building was sited only after extensive shadow study and analysis of the surrounding environment were performed. The results convincingly demonstrate the designer's artistry, sensitivity to the environment, and integrative role.

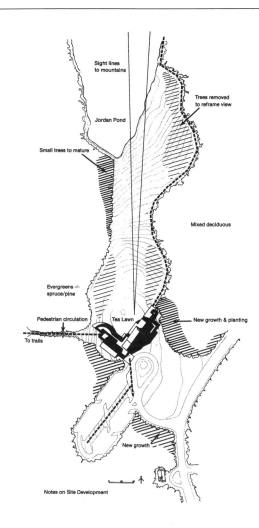

Sketch of site development showing orientation of Jordan Pond House. (Drawing courtesy Woo & Williams)

Interior of Jordon Pond House includes displays of historic artifacts. (Photo courtesy Woo & Williams)

Original view of rebuilt Jordan Pond House is retained by protection of open space. (Photo courtesy Woo & Williams)

Ponderosa Lodge, Mount Hermon, California

Johnson Johnson & Roy/inc., Landscape Architects

Ponderosa Lodge demonstrates an astute and ecologically sound design solution to a difficult task—placing a young adult and teenager religious conference facility on a fragile mountainside. Close collaboration convinced the owner to eliminate several intensive land use aspects of the program and protect much of the site as a nature preserve. The natural environment of the site dictated the size and intensity of development. The lodge was carefully located on a gentle slope, where intensive use has least impact. Automobiles are stored at the perimeter of the site, so that the natural beauty is in no way diminished. Views of the Pacific Ocean, canopies of live oak and madrone trees, and dense shade from redwoods were assets of considerable relevance to the siting of all structures. All building design is sensitive to the rustic setting.

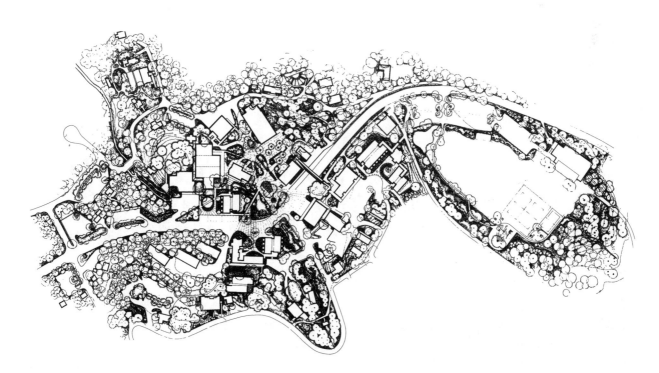

Plan of Ponderosa Lodge conference area development, with buildings sited to protect landscape assets. (Drawing courtesy Johnson Johnson & Roy/inc.)

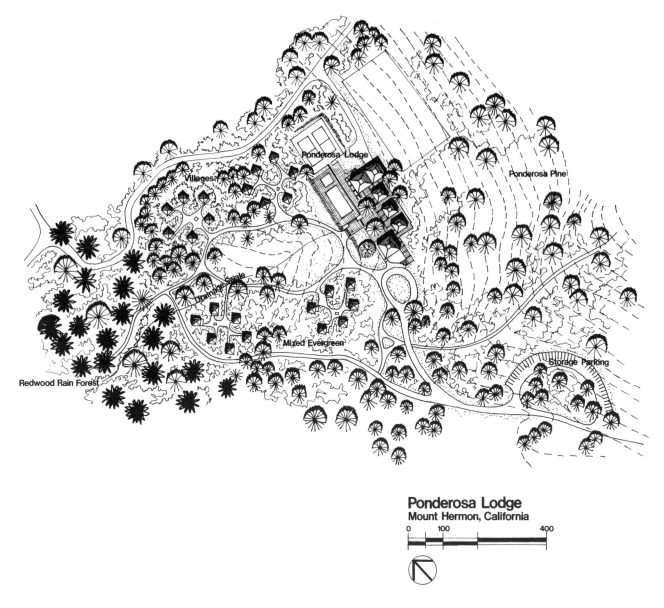

Ponderosa Lodge
Mount Hermon, California

0   100   400

Plan of Ponderosa Lodge site. (Drawing courtesy Johnson Johnson & Roy/inc.)

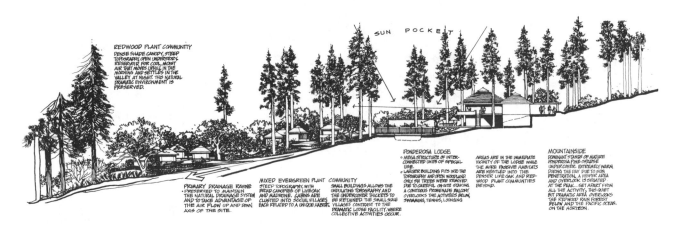

One of many studies made to make sure environmental assets were protected at Ponderosa Lodge. (Drawing courtesy Johnson Johnson & Roy/inc.)

Sketch of Ponderosa Lodge, used for communication with client. (Drawing courtesy Johnson Johnson & Roy/inc.)

Hard surfaces (left) are kept to a minimum in all development at Ponderosa Lodge to protect the environment. (Photo courtesy Johnson Johnson & Roy/inc.)

View of completed Ponderosa Lodge (right) showing unusual protection of natural resources. (Photo courtesy Johnson Johnson & Roy/inc.)

Fort Mackinac, Mackinac Island, Michigan

Eugene T. Petersen, Historic Restorer

One of the first restorations not only to contain rebuilt authentic structures and historic artifacts but to provide dramatizations and interpretation for visitors was Fort Mackinac, on a small island at the tip of Lake Huron. Thousands of visitors are attracted to the island annually because of its abundance of turn-of-the-century resort buildings, including the magnificent Grand Hotel, and because of its interesting nature and historic trails. The fort, with dramatizations of historic events, is the central feature of the destination. To avoid logistic problems of shipping and parking tourists' cars and to preserve the island's historic atmosphere, all cars are prohibited.

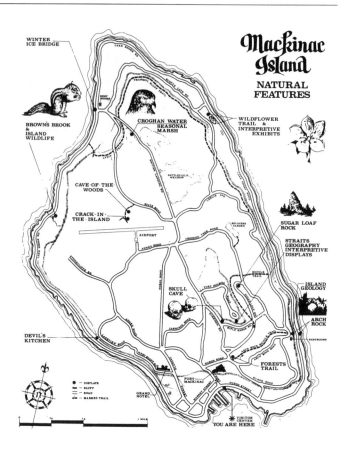

This 4-square-mile (10-square-kilometer) island where Fort Mackinac is located played an important role, under French, British, and American flags, in the settlement of North America. (Drawing courtesy Mackinac Island State Park Commission)

Uniformed guides help interpret the many events of historic Fort Mackinac, now restored. (Photo courtesy Mackinac Island State Park Commission)

Signers Memorial, Washington, District of Columbia

EDAW, Landscape Architects

State and national shrines, such as this memorial to the signers of the Declaration of Independence, are of increasing significance to travelers. The designers here sought to achieve a contemplative feeling to complement the plants and ponds of Constitution Gardens. To prevent it from competing with the tall memorials that surround it, the designers kept this memorial low and horizontal. This decision reflected both close cooperation between the designers and the client (the U.S. National Park Service) and sensitivity to visitors.

The design of the Signers Memorial (right), which was intentionally kept low, offers a dignified and harmonious solution for a national visitor shrine. (Photo courtesy Maxwell MacKenzie)

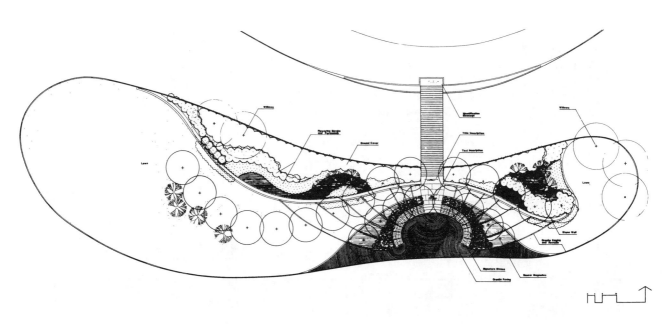

The finely designed Signers Memorial, constructed from fifty-six pink granite blocks, is neat and classical. (Drawing courtesy EDAW)

Tom McCall Waterfront Park, Portland, Oregon

Mitchell Nelson Group, Inc., Landscape Architects

This solution superbly meets the challenge of designing a multipurpose park in a run-down urban core. As an award jury commented, this area is now "a spacious, sufficiently formal, park-like area, uncluttered by constructions and ingenious walls, fountains, steps, etc. Serves everyday leisure activities and certainly increases the dignity of the city." Visitors and residents can now enjoy repose, exciting festivals, or boat access in the very heart of the city.

Open entertainment near water's edge at Waterfront Park. (Photo courtesy James Lemkin)

Waterfront Park offers beauty and boat access in the heart of a large city. (Photo courtesy James Lemkin)

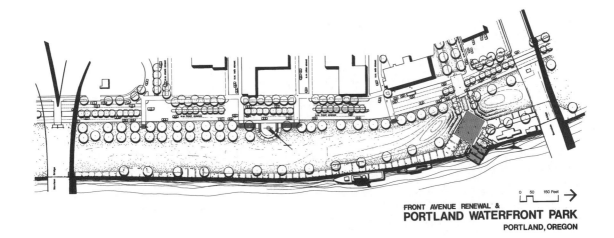

FRONT AVENUE RENEWAL &
PORTLAND WATERFRONT PARK
PORTLAND, OREGON

Plan of Portland Waterfront Park showing park and boat access (at right). (Plan courtesy Mitchell Nelson Group, Inc.)

Jackson Brewery Development, New Orleans, Louisiana

Concordia, Architects; Landesign, Landscape Architects

The renovated Jackson Brewery is an excellent example of historic reuse. It is part of the attraction complex of the French Quarter, within the destination zone of New Orleans and vicinity. Originally built in 1891, this landmark was reopened in 1984 after having been closed for ten years. The design challenge, well met, was to comply with the stipulations of the U.S. Department of the Interior for historic restoration, with the regulations of the Vieux Carré Commission (the local preservation body), and with the needs of visitors to a shopping-entertainment complex. The architecture reflects the Romanesque origins and yet sparkles with contemporary appeal. This festive redevelopment contains six floors of shops, restaurants, and craft stores, as well as spectacular views of the Mississippi River waterfront and the French Quarter.

First-phase redevelopment of the Jackson Brewery, adjacent to Jackson Square, French Quarter, New Orleans. (Photo courtesy Jackson Brewery Development Corporation)

Sketch of market complex attraction developed from defunct Jackson Brewery. (Drawing courtesy Jackson Brewery Development Corporation)

The Waterside, Norfolk, Virginia

*Wallace Roberts and Todd, Architects,
Landscape Architects*

This project illustrates the integrative role of designers. The architects and landscape architects involved first worked with the city of Norfolk to create the Downtown 1990 Plan, which provided long-range guidance for revitalization and future development for visitors as well as residents. Then, several projects, including the Waterside, were planned for collaborative public and private development of this busy waterfront. The Waterside is a pleasant shopping, eating, and entertainment attraction. It includes space and facilities for five restaurants, twenty-two fast food enterprises, thirty-five specialty retail and market produce shops, thirty-four kiosks, and eighteen pushcart vendors.

Downtown Norfolk 1990 Plan identifying overall projected revitalization of the entire area. (Drawing courtesy Wallace Roberts and Todd)

Creative redevelopment of harbor waterscape and architecture in Norfolk has added vitality and purpose for visitors and residents. (Photo by Bruce A. Tamte, courtesy Wallace Roberts and Todd)

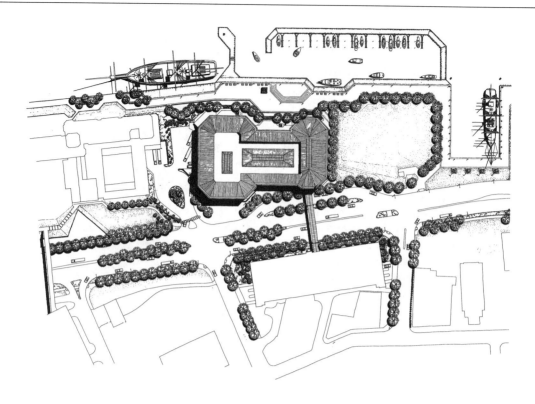

Detail plan of Waterside project showing relationship of harbor and waterfront structures. (Drawing courtesy Wallace Roberts and Todd)

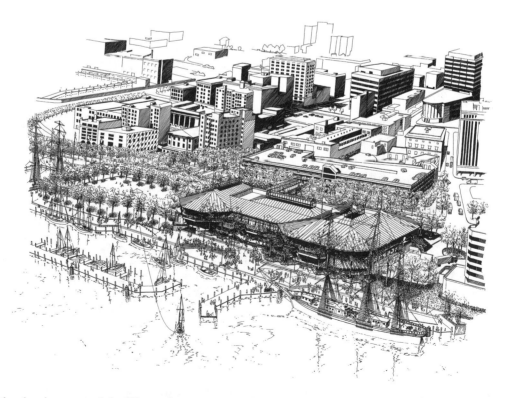

Sketch plan of redevelopment of the Waterside project area. (Drawing courtesy Wallace Roberts and Todd)

Clarke Quay, Singapore

*ELS/Elbasini and Logan and RSP,
Singapore, Architects; EDAW, Landscape
Architects*

The design challenge was to create a five-block attraction including abandoned historic warehouses fronting on the Singapore River. Included in the area were a central square, shade trees, a tropical garden, an aviary, landscape paving, shelters, fountains, and lighting. The resulting design integrates historic restoration mixed with contemporary commercial and entertainment facilities.

Waterfront view of shops and historic restoration at Clarke Quay. (Photo courtesy EDAW)

Layout plan of Clarke Quay adaptive reuse project. (Drawing courtesy EDAW)

Puerto Azul, Philippines

Klages Carter Vail, Architect; EDAW, Landscape Architects

Creation of a new community for residents and visitors on Manila Bay without destruction of the character of the land was the challenge for this project. Initial stages of development included recreation areas, the lagoon beach, a golf course and academy, three hotels, an equestrian center, townhouses, villas, and condominiums. All new construction had to take place in sustainable settings. Environmental design policies included retention of undisturbed watersheds, open space buffer areas, enhanced wetland habitats, and lagoons for runoff control. More than half of the site is preserved forever for open space and scenic appeal.

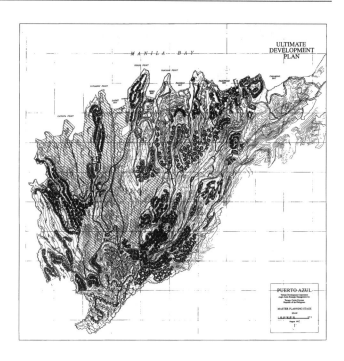

Master planning stage for Puerto Azul. (Photo courtesy EDAW)

Perspective sketch of waterfront development at Puerto Azul. (Sketch courtesy EDAW)

Westonbirt Visitor Centre, Gloucestershire, England

Andris Berzins & Associates, Architects

Adapting to greater volumes of tourists, designers are creating functional and attractive visitor centers appropriate to each setting. This structure-and-landscape complex, a focal point for an entire arboretum, is an organic design solution for a dominantly vegetative setting. The main pavilion contains flexible exhibition space, a seminar/audiovisual room, a library, and offices for staff. The adjacent pavilion forms a courtyard, provides refreshments, and is rotated 45 degrees from the main pavilion to create design tension between the two forms. The structures relate superbly to the site, a 150-year-old arboretum containing two thousand trees of fifty-five species.

Plan showing environmental orientation of Westonbirt Visitor Centre structures (below). (Drawing courtesy Forestry Commission)

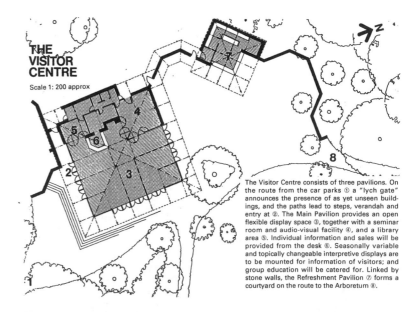

THE VISITOR CENTRE

Scale 1: 200 approx

The Visitor Centre consists of three pavilions. On the route from the car parks ① a "lych gate" announces the presence of as yet unseen buildings, and the paths lead to steps, verandah and entry at ②. The Main Pavilion provides an open flexible display space ③, together with a seminar room and audio-visual facility ④, and a library area ⑤. Individual information and sales will be provided from the desk ⑥. Seasonally variable and topically changeable interpretive displays are to be mounted for information of visitors; and group education will be catered for. Linked by stone walls, the Refreshment Pavilion ⑦ forms a courtyard on the route to the Arboretum ⑧.

The design of Westonbirt Visitor Centre provides necessary functions and yet blends well with the setting (below). (Photo courtesy Forestry Commission)

Punakaiki Visitor Centre, Greymouth, New Zealand

Gary Hopkinson & Associates, Architects

Because of its compatibility with the surrounding environment, this design has received many awards. The service center, which contains toilets, a kitchen, and recreational facilities, supplements the adjoining Dolomite Scenic Reserve, a major attraction of the region. Nearby, in the Punakaiki Information Centre, displays, information, and an audiovisual presentation are integrated with a tea room and souvenir shop for a functional attraction complex.

Simple, straightforward design of Punakaiki Information Centre. (Photo courtesy Gary Hopkinson & Associates)

Harmonious camping facility near Punakaiki Visitor Centre. (Photo courtesy Gary Hopkinson & Associates)

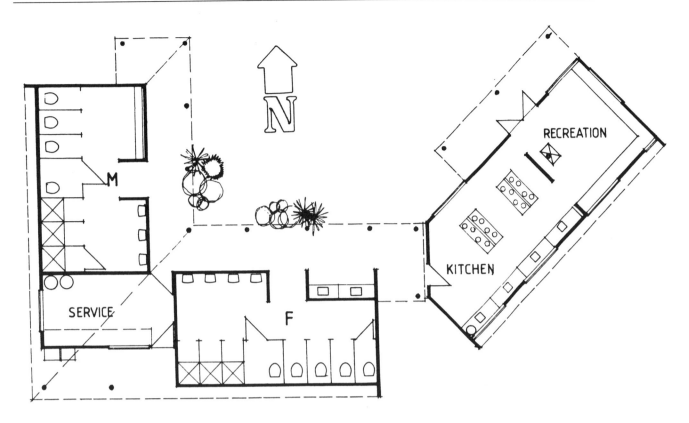

Plan of Punakaiki Service Center for Dolomite Scenic Reserve. (Drawing courtesy Gary Hopkinson & Associates)

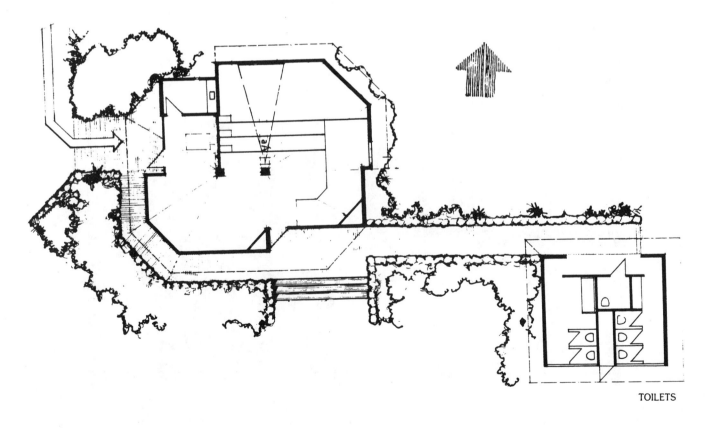

Punakaiki Information Centre plan. (Drawing courtesy Gary Hopkinson & Associates)

Giant's Causeway Centre, Belfast, Northern Ireland

Andris Berzins, Architect; Robin Wade, Interpretive Designer

This informative center provides a meaningful experience of the Giant's Causeway, a significant tourist attraction along Ireland's north coast, often called the Eighth Wonder of the World. The exterior design is appropriate to the barren basaltic headlands. The creative interior includes explanations and exhibits of the area's geology, personalized information, retail sales of literature and mementos, and descriptive audiovisual presentations.

Interior of the Giant's Causeway Centre includes space for souvenir sales as well as educational exhibits. (Photo courtesy Andris Berzins)

Entrance approach to Giant's Causeway Centre. (Photo courtesy Andris Berzins)

Detail view of unusual geological spectacle at Giant's Causeway on Irish coast. (Photo courtesy Andris Berzins)

Mary Rose Tudor Ship Museum, Portsmouth, England

Andris Berzins & Associates, Architects

A major repair, restoration, and conversion of an ancient timber-framed boathouse monument has produced an interesting focal point for visitors. This museum displays a sampling of the seventeen thousand artifacts recovered from the *Mary Rose*, a ship carrying seven hundred men and ninety-one guns that was ordered by King Henry VIII to play a key role in repelling a French invasion. Unfortunately for the English, as the ship set sail from Portsmouth in 1545, it was capsized by a gust of wind, and all but thirty of its men were lost. Remains of the hull were raised in 1982.

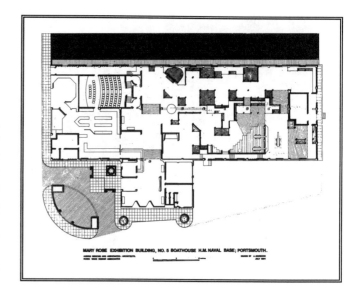

Plan of Mary Rose Tudor Ship Museum in ancient boathouse. (Drawing courtesy Andris Berzins)

View of museum interior with artifacts from the Mary Rose. *(Photo by Martin Charles, courtesy Andris Berzins)*

Mill Creek Restoration and Visitor Center, Mackinaw City, Michigan

Victor Hogg, Interpretive Designer

After intensive land analysis and archaeological study, the 1780 sawmill near Mackinaw City was rebuilt on its original site. Reconstructed houses, a dam, a fish ladder, and an Indian camp round out the display. In the visitor center, interpretation of natural resources and how mankind has used them from the Ice Age to the present offer tourists an educational as well as an entertaining experience.

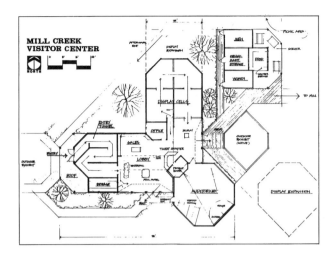

Plan of Mill Creek Visitor Center, housing displays and other interpretive aids. (Photo courtesy Mackinac Island State Park Commission)

Sketch plan of Mill Creek restoration complex, simulating the 1780 structures. (Drawing by Victor R. Nelhiebel; courtesy of Mackinac Island State Park Commission)

Visitors to Mill Creek Visitor Center gain an understanding of how this early mill was located and operated. (Photo courtesy Mackinac Island State Park Commission)

Visitors to Mill Creek learning about mill operations from an interpreter. (Photo courtesy Mackinac Island State Park Commission)

PGA National Resort, Palm Beach Gardens, Florida

Schwab & Twitty, Inc., Architects; Urban Design, Inc., Landscape Architects

This 4-square-mile destination complex consists of several types of housing and services surrounding four championship golf courses. Responding to a market trend of retirees seeking golf, this design maximizes exposure to freeway frontage yet provides ample privacy. Intensity of housing varies. The complex includes a hotel, restaurants, a swimming pool, tennis courts, a fitness center, and a dance studio. A series of lakes, canals, and weir structures protect groundwater levels to minimize salt water intrusion. Housing design in separate neighborhoods varies, but the golf course setting retains harmony throughout.

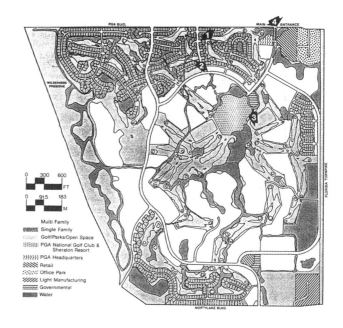

Plan for major destination complex at PGA National Resort. (Drawing courtesy Urban Design, Inc.)

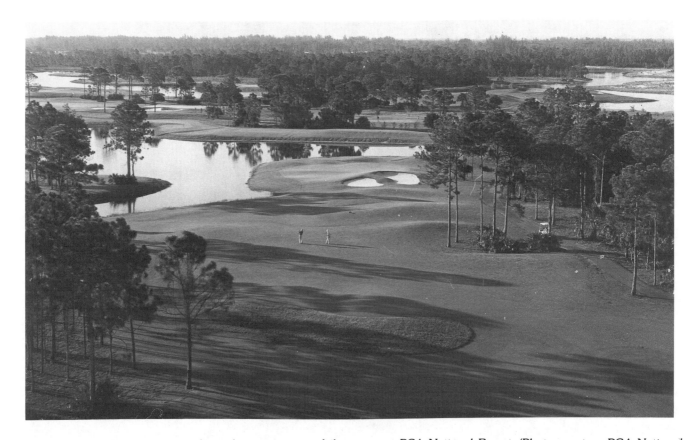

Four championship golf courses form the centerpiece of the resort at PGA National Resort. (Photo courtesy PGA National Resort)

Many recreation opportunities are available at PGA National Resort, including swimming. (Photo courtesy PGA National Resort)

Mauna Lani Resort, South Kohala, Hawaii

Belt, Collins & Associates, Landscape Architects

This outstanding, award-winning design exemplifies the fulfillment of all tourism development goals: it stimulates business, is sensitive to the environment, and meets market needs. Because of their archaeological and historical value, prehistoric Kalahuipua'a fish ponds and historic sites have been kept in preserves. The 351-room Mauna Lani Bay Hotel and private homes (at Mauna Lani Terrace) have been sited to protect these preserves and to maintain aesthetic vistas to the ocean and mountains. The area also includes the well-designed Francis H. I'i Brown Golf Course. The rugged natural beauty of the volcanic landscape has been retained with skill. New native plantings, as well as a six-story atrium within the hotel, enhance the site.

Site plan for the Mauna Lani Bay Hotel (right) is sensitive to land assets of topography and vistas to the ocean. (Drawing courtesy Belt, Collins & Associates)

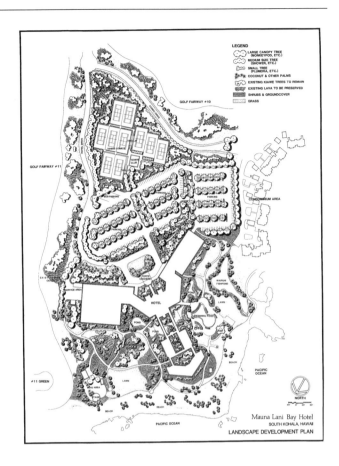

View of completed Mauna Lani Bay Hotel reveals design creativity in both orientation and utilization of site. (Photo courtesy Belt, Collins & Associates)

Parador Naçional de Segovia, Segovia, Spain

Secretariat of State for Tourism, Developers

This new hotel was built several kilometers from Segovia to avoid aesthetic intrusion upon the architecture and landscape of that ancient city, occupied by the Romans in about 100 B.C. The architectural style is fairly contemporary but retains the red-brick color and general feeling of the old city. The lobby presents a framed view of the city.

With tasteful use of materials that simulate the old city of Segovia, the new hotel design is quite compatible with its theme. (Photo courtesy Secretariat of State for Tourism, Spain)

View of Segovia from newly built hotel, intentionally located well outside the ancient city.

Parador Naçional de Siguenza, Siguenza, Spain

Secretariat of State for Tourism, Developers

Few countries have engaged in adaptive reuse of historic structures for tourism better than Spain, with its many *paradores*. These inns are converted castles, palaces, and convents that retain the basic traditions of past centuries but have modern hotel conveniences. An example is the Parador Naçional "Castilla de Siguenza," located 128 kilometers from Madrid and converted to a large hotel in the 1970s. *Paradores,* which are located in small towns or rural areas throughout the country, offer a rich travel experience not available in the more standardized hotels of Spain's large cities.

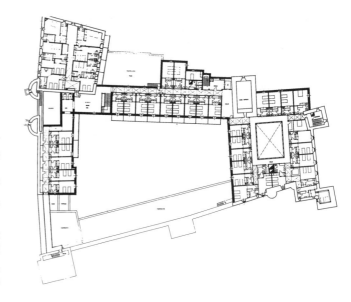

Plan showing remodeling of second floor into guest rooms at Parador Naçional de Siguenza. (Drawing courtesy Secretariat of State for Tourism, Spain)

The ancient Castilla de Siguenza is now converted to tourist lodging. (Photo courtesy Secretariat of State for Tourism, Spain)

Interior courtyard (right) of the old castle at Siguenza has been modified very little. (Photo courtesy Secretariat of State for Tourism, Spain)

Jupiters Casino, Broadbeach Island, Queensland, Australia

Sprankle, Lynd & Sprague, Architects

This exceptional design complex, intimately linked with a 622-room Conrad Hilton International Hotel, illustrates travel services integrated with an attraction complex. The interior design not only is sensitive to activities within but maximizes sweeping views of the gardens, the Pacific Ocean, and hinterland mountain ranges. The complex includes a wide array of attractions, such as a discotheque, a convention center, tennis courts, a swimming complex, and outdoor entertainment areas.

A garden atrium forms a link between the hotel and casino. A translucent canopy ties the rectangular hotel design to the boomerang-shaped casino. (Photo by Chris Pilz, courtesy Sprankle, Lynd & Sprague)

Brightleaf Square, Durham, North Carolina

Ferebee, Walters & Associates, Architects

Established downtowns can compete well with glitzy suburban shopping centers, particularly where there is creative redesign, new merchandising, and a mixture of tourists and residents. Originally built as tobacco factories, Brightleaf Square now retains the patina of an earlier era but provides intimate and distinctive shops, restaurants, and offices. Tourists can rediscover the warmth and pleasures of urban destinations when they are designed with such feeling as this.

View of redeveloped Brightleaf Square. (Photo courtesy Ferebee, Walters & Associates)

"Before" view of the blighted Brightleaf site. (Photo courtesy Ferebee, Walters & Associates)

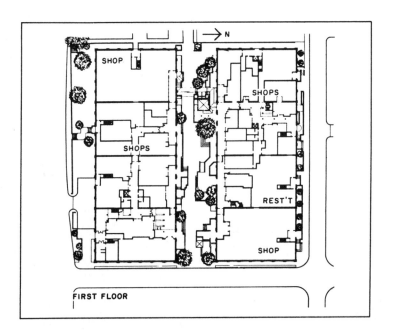

Plan of Brightleaf Square's restored tobacco factories. (Drawing courtesy Ferebee, Walters & Associates)

Station Square, Pittsburgh, Pennsylvania

Pittsburgh History and Landmarks Foundation, Developers

Part of a long-range revitalization of the Monongahela riverfront, situated across from Pittsburgh's downtown core, Station Square is a 41-acre tract with obsolete and underutilized buildings dating from 1897 to 1917, owned by the Pittsburgh & Lake Erie Railroad. The old waiting room has been converted to a dining area; the old baggage room to a seafood and cocktail bar; the old dining room and ladies' waiting room to new dining areas; and an outdoor walkway to a dining area with a view of the downtown. Signage reflects the railroad theme. The entire complex is designed for maximum historic education and use.

One of several dining areas at Station Square (right), providing romantic and nostalgic atmosphere for visitors in a once-blighted area. (Photo courtesy Pittsburgh History and Landmarks Foundation)

Festive visitors enjoy the renovated Station Square area, complete with restored freight house on the right. (Photo courtesy Pittsburgh History and Landmarks Foundation)

Albert Street Revitalization, Winnipeg, Manitoba

Hilderman Witty Crosby Hanna & Associates, Landscape Architects

The first streetscaping project in the Old Market Square District of Winnipeg, Albert Street has set the character for the entire district's restoration. Street parking continues, but new pedestrian bays reach into former travelways, increasing the attractiveness of the area and the efficiency with which people use it. The designer has used interlocking pavers, concrete headers, brick-and-trim half circles, antique-styled bollards, benches, and simulated period light standards to complement the restoration of the historic building facades. The City Council Committee on the Environment coordinates design for all redevelopment in the district.

Renovated hotel entrance at Albert Street (right) illustrates design details of paving, curb, bollards, and planters. (Photo courtesy Hilderman Witty Crosby Hanna & Associates)

General view of Market Square District shows planters, pedestrian plaza, and period-style lamp standards. (Photo courtesy Hilderman Witty Crosby Hanna & Associates)

Durango–Silverton Narrow Gauge Railroad, Colorado

*Durango–Silverton Narrow Gauge R. R.,
Developers*

More than just a trip back in time, this 45-mile restoration is a true attraction complex. It includes spectacular mountain scenery, old gold mines, diverse geological formations, and the flora and fauna of several life zones, as well as relics of old trains. The towns of Durango, Rockwood, Needleton, and Silverton are within the railway corridor and are representative of this frontier mining territory. Silverton and the Durango–Silverton narrow gauge D&RGW railroad have been designated national historic landmarks by the U.S. National Park Service.

An example of a turn-of-the-century steam train attraction, the Durango–Silverton railroad passes through spectacular landscapes. (Photo courtesy Durango & Silverton Narrow Gauge RR)

Fiddler's Green Amphitheatre, Englewood, Colorado

Hargreaves Associates, Landscape Architects

An award jury commented on this project, "Part of a new process of educating clientele to use projects to make connection with the community. Done with verve and gusto. Magic making—a special place for special happenings" (Fiddler's Green Amphitheatre, 1984, 66). An open theater for the performing arts, this dramatic landscape attraction has promise of becoming a destination point for visitors as well as residents.

Festival performance at Fiddler's Green in process, with innovative stage backdrop. (Photo courtesy Hargreaves Associates)

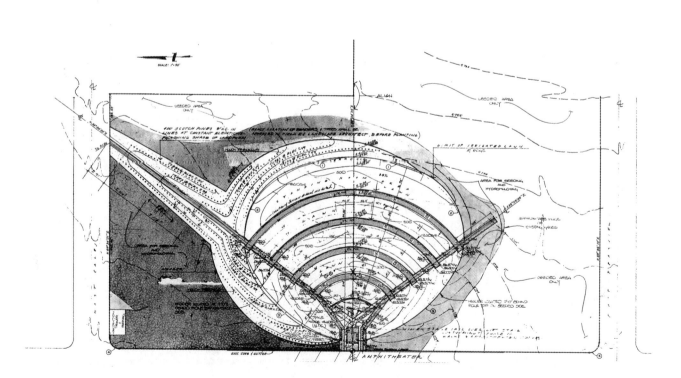

Plan of Fiddler's Green (below), an unusual outdoor theater, indicating contours for reshaping the land. (Drawing courtesy Hargreaves Associates)

Scenic Tour Barge, Paris, France

Salt and Pepper Tours, Developers

Waterways are increasingly popular for tourism because they offer an experience unequalled by any other attraction form. Simply by fitting out barges to accommodate visitors, designers enable small groups to be enriched by the waterfront landscape of rivers and canals throughout Europe at low cost.

Casual interior of a scenic tour barge (below) provides for private relaxing as well as viewing of the scenic river corridor. (Photo courtesy Salt and Pepper Tours)

La Patache Autobus *(tour barge)* on the St. Martin Canal, Paris. *(Photo courtesy Salt and Pepper Tours)*

Infomart, Dallas, Texas

Growald & Associates, Architects

Called the world's first information processing market center, this building with its innovative design allows many computer and telecommunications manufacturers to display their products. As a center for displays, meetings, conferences, and receptions, it provides an important information link for trade in normal retail outlets. Reflecting the increased sophistication of business and trade travel attractions, the building was inspired by London's Crystal Palace and contains 1.5 million square feet of space.

Interior view of central reception lobby of Infomart, leading to market areas. (Photo courtesy Dallas Market Center Company)

Elevation drawing of Infomart facade, reminiscent of London's Crystal Palace. (Drawing courtesy Dallas Market Center Company)

Finished Infomart building, housing an entirely new concept in marketing. (Photo courtesy Dallas Market Center Company)

Tree House, New Children's Zoo, Philadelphia, Pennsylvania

Venturi, Rauch, and Scott Brown, Architects

A creative design for a traditional zoo, Tree House adds fantasy to plant and animal forms. This award-winning project appeals to adults as well as children and to visitors as well as residents. It utilizes a 111-year-old Victorian structure, providing a meaningful marriage between revered architecture and the wonders of science. The freedom from signs allows the visitors to experience the animals and plants, to learn about them without distraction.

Exterior of century-old zoo structure, now converted to Tree House. (Photo courtesy Venturi, Rauch, and Scott Brown)

Entrance with vaulted ribs sets the theme for the fantasy world to come in Tree House. (Photo by Matt Wargo; courtesy Venturi, Rauch, and Scott Brown)

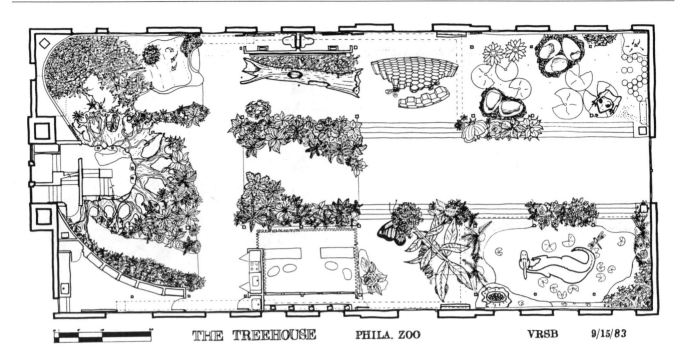

THE TREEHOUSE   PHILA. ZOO   VRSB   9/15/83

Plan of Tree House interior complete with swamp, everglade, beaver pond, and rainforest. (Drawing courtesy Venturi, Rauch, and Scott Brown)

Designs made of plastic in Tree House are hybrids of sculpture, architecture, and furniture, offering a stimulating animal playground atmosphere. (Photo by Tom Crane; courtesy Venturi, Rauch, and Scott Brown)

Zululand Tree Lodge, Maputaland, South Africa

Ridler Shepherd Low, Architects

Designed within the theme of ecotourism, this resort and conference retreat is in a wildland setting and yet readily accessible. It is located near the main north–south road between Mpumalanga and Durban and an airport at Richards Bay. Several chalets are linked by convenient walkway to the reception lounge, central dining lounge, and conference room. The development process of design and construction involved more than 200 local residents as well as the designers. Construction materials and methods such as the chalets' elevation on stilts ensure an environmentally sound as well as aesthetically pleasing development. The site chosen protects major indigenous assets and yet offers visitors views of native plants and animals centered on the valley of the Mzinene River.

A typical treehouse at Zululand Tree Lodge, illustrating sensitive design adaptation to the African setting. Built on stilts, the chalets create minimal environmental disturbance. (Courtesy Southern Sun Group)

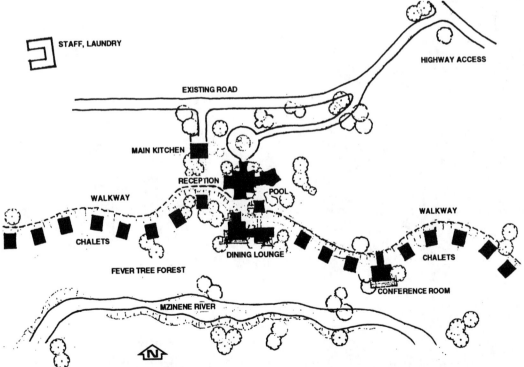

The site plan for the Zululand Tree Lodge emphasizes chalet privacy and proximity to indigenous resources as well as modern accommodations, services, and conference facilities for a high-level travel market. (Plan by Landmark Studios, courtesy Southern Sun Group)

Phinda Forest Lodge, Mkuzi Game Reserve, South Africa

The Conservation Corporation, Maresch Ridler Shepherd, Architects

Built by local labor and with community involvement in design decisions, this thirty-two-bed facility demonstrates great environmental sensitivity and response to a special natural resource travel demand. Structures were sited without disturbing existing trees and were built on stilts to protect forest soils. Tourists are able to view and photograph cheetahs, leopards, lions and a great diversity of bird life. Guided walking tours and boat cruises offer visitors a comprehensive African experience in the 17,500-hectare Phinda Resource Reserve and Mkuzi Game Reserve.

A typical guest room interior in the Phinda Forest Lodge illustrates the intimate design linkage between the outdoor natural setting and visitor comfort. (Courtesy Conservation Corporation)

Bibliography

CHAPTER 1

Tourism: Positive and Negative

Ashworth, G. L., and J. E. Tunbridge. 1990. *The tourist-historic city*. London: The Belhaven Press.

Brancatelli, Joe. 1995. What is polluting our beaches?" *Smithsonian*, April, 74–75.

Bronowski, Jacob. 1973. *The ascent of man*. Boston: Little, Brown & Co.

Butler, R. W. 1980. The concept of a tourist area cycle of evolution: Implications for management of resources. *Canadian Geographer* 24:1, 5–12.

Casellati, Antonio. 1991. Averting a tourism disaster. *MCL Newsletter,* March/December, 36–38.

Center for Marine Conservation. 1996. *1995 International coastal cleanup, U.S. results*. Texas section. Washington, DC: Center for Marine Conservation.

D'Amore, Louis J. 1988. Tourism—The world's peace industry. In *Proceedings of Tourism—A vital force for peace,* Montreal, 7–14.

Getz, Donald. 1994. In pursuit of the quality tourist. Paper presented at conference, Tourism Down Under, x–xx December 1994, at Massey University, Palmerston North, New Zealand.

Guiding principles of sustainable design. 1993. Denver: Denver Service Center, National Park Service.

Hawkins, Rebecca and Victor Middleton. 1993. The environmental practices and programs of travel and tourism companies. *World Travel and Tourism Review,* 3:163–171. Oxon, UK:CAB International.

Heaney, Colin. 1993. Preserving our lifestyle. Insert in *The Advocate* (Byron Shire, Australia): 30 June.

Hibbert, Christopher. 1969. *The grand tour*. New York: G. P. Putnam's Sons.

Hollinshead, Keith. 1993. Ethnocentrism in tourism. In *VNR's encyclopedia of hospitality and tourism,* eds. Mahmood Khan, Michael Olsen, and Turgut Var, 652–662. New York: Van Nostrand Reinhold.

Hollinshead, Keith. 1996. Marketing and metaphysical realism: The disidentification of Aboriginal life and traditions through tourism. In *Tourism and indigenous peoples,* ed. Richard Butler and Thomas Hinch, 308–348. London: International Thomson Business Press.

Koth, Barbara, Glenn Kraeg, John Sem, and Kathy Kjolhaug. 1991. Appendix A: Sandpoint, Idaho. In *A training guide for rural tourism development,* 10–14. St. Paul: University of Minnesota Extension Service.

Lea, John. 1988. *Tourism and development in the Third World*. London: Routledge.

Lennard, Suzanne, and Henry L. Lennard. 1995. *Livable cities observed*. Carmel, CA: International Making Cities Livable Council.

Lowenthal, David, and Hugh C. Prince. 1955. English landscape tastes. *Geographical Review* 55: 2, 82.

Miossec, J. M. 1977. Un modele de l'espace touristique. *L'Espace Geographique* 6: 1, 41–48.

Pearce, Douglas. 1989. *Tourist development*. 2d ed. New York: Longman (copublished with John Wiley & Sons).

Rees, W. E. 1989. Defining sustainable development. *CHS Research Bulletin,* University of British Columbia, 3 May.

Ritchie, J. R. Brent. 1991. Sustainable development in tourism: A framework for policy; an agenda for action. In *Tourism, environment, sustainable development: An agenda for research,* ed. Laurel J. Reid, 94–99. Location: Travel and Tourism Research Association Canada.

Roche, Maurice. 1994. Mega-events and urban policy. *Annals of Tourism Research* 21: 1–19.

Schapiro, Mark. 1995. Providencia: Battle for the soul of an island. *Conde Nast Traveler,* December.

Schmitt, Adolf. 1986. Cairo seminar summary. In *IFLA Yearbook 1985/86,* 154–159. Versailles, France: International Federation of Landscape Architects.

Selzano, Edoardo. 1991. The need for balance. *MCL Newsletter*, March/December, 35–36.

Smith, Valene L. 1977. *Hosts and guests—The anthropology of tourism.* Philadelphia: University of Pennsylvania Press.

Stanton, Max E. 1977. The Polynesian Culture Center: A multi-ethnic model of seven Pacific cultures. In *Hosts and guests,* ed. Valene L. Smith, 193–206. Philadelphia:, University of Pennsylvania Press.

Strutin, Michele. 1994. The perfect arena. *Landscape Architecture* 84: 1, 76–78.

Taylor, Gordon D. 1991. Tourism and sustainability—Impossible dream or essential objective. In *Tourism, environment, sustainable development: An agenda for research,* ed. Laurel J. Reid, 27–29. Location: Travel and Tourism Research Association Canada.

Texas Tourism Division. 1995. *Texas travel facts.* Austin: Texas Department of Commerce.

Tourism Stream Action Strategy Committee. 1990. *An action strategy for sustainable tourism development.* Ottawa: Tourism Canada.

Troyer, Warner. 1992. *The green partnership guide.* Toronto: Canadian Pacific Resorts.

Weaver, Glenn. 1991. *Tourism USA,* 3d. ed. Washington, DC: U.S. Travel and Tourism Administration.

Werkmeister, Hans Friedrich. 1986. The perfect way to Tut-Ank-Amun. *IFLA Yearbook 1985/86,* 182–184. Versailles, France: International Federation of Landscape Architects.

Wight, Pamela. 1994. The greening of the hospitality industry: Economic and environmental good sense. In *Tourism: State of the art,* eds. A.V. Seaton, R. Wood, P. Dieke, and C.L. Jenkins, 665–674. Chichester, England: John Wiley & Sons.

World Travel & Tourism Council. 1993. *Travel & tourism: Environment & development.* Brussels, Belgium.

World Travel & Tourism Council. 1995. *WTTC Report (1995).* Brussels, Belgium:World Travel & Tourism Council.

Xiang, Sun Xiao. 1995. Tourism development and its gains and losses on natural, historical environments and culture in modern China. In *Proceedings of the 32nd IFLA World Congress,* 56–59. Meeting of the International Federation of Landscape Architects. Bangkok: Thai Association of Landscape Architects.

CHAPTER 2

Politics and Ethics

Alberta Tourism Partnership. 1996. *Uplink.* Issue 1, Spring. Edmonton: Alberta Tourism Partnership.

Andereck, Kathleen L. 1995. Arizona's rural tourism development program. *Women in Natural Resources* 16: 8–11.

ARA Consulting Group. 1991. *Developing a code of ethics: British Columbia's tourism industry.* Victoria, British Columbia: British Columbia Ministry of Development, Trade and Industry.

Ashworth, G. J., and J. E. Tunbridge. 1990. *The tourist-historic city.* London: Belhaven Press.

Canada. Ministers of Culture and Tourism and Recreation Advisory Committee. 1994. *Ontario's tourism industry: Opportunity, progress, innovation.* Ottawa: Queen's Printer.

Darrow, Kit. 1995. A partnership model for nature tourism in the eastern Caribbean islands. *Annals of Tourism Research* 22: 48–51.

Dieke, Peter. 1993. Tourism and development policy in the Gambia. *Annals of Tourism Research* 20(3): 423–449.

Edgell, David L. 1990a. *Charting a course for international tourism in the nineties.* Washington, DC: U.S. Travel and Tourism Administration.

Edgell, David L. 1990b. *International tourism policy.* New York: Van Nostrand Reinhold.

Edgell, David L. 1993. *World tourism at the millennium.* Washington, DC: U.S. Travel and Tourism Administration.

Giltmier, James W. 1991. Historical perspectives on rural land transitions. In *Enhancing rural economies through amenity resources,* 37–41. State College: Pennsylvania State University.

Hall, Colin Michael. 1994. *Tourism and politics.* New York: John Wiley & Sons.

Hawkins, Ann E. 1994. The Ontario Tourism Council: A common ground for industry. In *TTRA Canada conference proceedings,* 116–123. Ottawa: Travel & Tourism Research Association Canada.

Howe, Margherita. 1994. The good, bad and the ugly. In *TTRA Canada conference proceedings.* Ottawa: Travel & Tourism Research Association Canada.

Jamal, Tazim B., and Donald Getz. 1995. Collaboration theory and community tourism planning. *Annals of Tourism Research* 22(1): 186–204.

King, Brian, Abraham Pizam, and Ady Milman. 1993. Social impacts of tourism: Host perceptions. *Annals of Tourism Research* 20: 650–665.

Lankford, Samuel V. 1994. Attitudes and perceptions toward tourism and rural regional development. *Annals of Tourism Research* 32(3): 35–43.

Lawson, John. 1991. Who is responsible for achieving sustainable development? In *Tourism, environment, sustainable development: An agenda for research,* ed. Laurel J. Reid, 105–106. Ottawa: Travel and Tourism Research Association Canada.

Lea, John. 1988. *Tourism and development in the Third World.* New York: Routledge.

Lea, John. 1993. Tourism development ethics in the Third World. *Annals of Tourism Research* 20: 707–715.

Leopold, Aldo. 1949. *Sand County almanac.* New York: Oxford University Press.

Leyva, Michael. 1996. Rural tourism development. In Highlights of the 27th annual TTRA conference, Las Vegas. *Journal of Travel Research* 35(1): 78–88.

Lipman, Geoffrey H. 1995. Taxing travel. *Viewpoint* 2: 52–60.

Mudrick, Gunter P. 1991. The European campaign for the countryside: A Council of Europe initiative for the future of women and men in the countryside. In *Enhancing rural economies through amenity resources,* 181–187. State College: Pennsylvania State University.

Nelson, J. Gordon. 1991. Are tourism growth and sustainability objectives compatible? In *Tourism, environment, sustainable development: An agenda for research,* ed. Laurel J. Reid. Ottawa: Travel & Tourism Research Association Canada.

O'Grady, Ron. 1980. *Third World tourism.* Singapore: Christian Conference of Asia.

Organization for Economic Cooperation and Development. 1990. *Tourism policy and international tourism in OECD member countries.* Paris: Organization for Economic Cooperation and Development.

Petulla, Joseph M. 1980. *American environmentalism.* College Station: Texas A&M University Press.

Pritchett, Victor S. 1964. *The offensive traveler.* New York: Alfred A. Knopf.

Przeclawski, Kizysztof. 1995. Deontology of tourism. Paper presented at conference of the International Academy for the Study of Tourism, 24–30 June, Cairo, Egypt.

Richter, Linda K. 1989. *The politics of tourism in Asia.* Honolulu: University of Hawaii Press.

Richter, Linda K. 1994. The political dimensions of tourism. In *Travel, tourism, and hospitality research.* 2d ed., eds. J. R. B. Ritchie and C. R. Goeldner, 219–231. New York: John Wiley & Sons.

Rumble, Neil. 1994. No sour grapes. In *TTRA Canada conference proceedings 1994.* Ottawa: Travel & Tourism Research Association Canada.

Scottish Tourist Board. n.d. *Making the directive work for you.* Edinburgh: Scottish Tourist Board.

Scottish Tourist Board. 1995a. *Corporate Plan 1995/6 to 1997/8* (1995). Edinburgh: Scottish Tourist Board.

Scottish Tourist Board. 1995b. *Development, objectives and functions 1995–1996.* Edinburgh: Scottish Tourist Board.

Shepard, Paul. 1967. *Man in the landscape.* New York: Alfred A. Knopf.

Smith, Valene L. 1995. The 4-H's of tribal tourism: Acoma, a Pueblo case study. Paper presented at the biennial meeting of International Academy for the Study of Tourism, 24 June–1 July, Cairo, Egypt.

Stauskas, Vladas. 1995. Tourism development and landscape protection. In *Proceedings of the 32nd IFLA World Congress,* 49. Meeting of the International Federation of Landscape Architects. Bangkok: Thai Association of Landscape Architects.

Stokes, Peter. 1994. Let's get it straight. In *TTRA Canada conference proceedings 1994.* Ottawa: Travel & Tourism Research Association Canada.

Stokowski, Patricia A. 1992. The Colorado gambling boom: An experiment in rural community development. *Small Town* 22(6), 12–19.

Thailand: Passport's illustrated guide. 1993. Lincolnwood, IL: Passport Books.

Tourism Alberta. 1991. *Tourism 2000: A vision for the future.* Edmonton: Tourism Alberta.

Upchurch, Randall S., and Sheila K. Rohland. 1995. An analysis of ethical work climate and leadership relationship in lodging operations. *Journal of Travel Research* 34: 36–42.

Western Australia Tourism Commission. 1989. *The eco-ethics of tourism development.* Perth: Western Australia Tourism Commission.

World Bank. 1993. *Environmental assessment sourcebook.* (1993) Washington, DC: The World Bank.

World Travel & Tourism Environmental Research Center. 1993. *World Travel & Tourism Environment Review 1993.* Headington, Oxford: World Travel & Tourism Environment Research Center.

WTTC. 1993. Bureaucratic barriers to travel. In *World Travel and Tourism Review.* Vol. 3, 90–93. Wallingford, Oxon: CAB International.

Wright, Michael, and Gary Stoller. 1995. America disconnected. *Conde Nast Traveler,* September.

CHAPTER 3

Tourism Function: Demand

Bruner, Jerome S. 1951. Personal dynamics and the process of perceiving. In *Perception,* eds. Blake and Ramsey. New York: The Ronald Press.

Cessford, G. R., and P. R. Dingwall. 1994. Tourism on New Zealand's sub-Antarctic islands. *Annals of Tourism Research* 21: 318–332.

Crompton, John L. 1979. Motivations for pleasure vacation. *Annals of Tourism Research* 6: 408–424.

Dychtwald, Ken. 1989. *The shifting American marketplace.* Emeryville, CA: Age Wave, Inc.

Edgell, David L., Sr. 1990. *International tourism policy.* New York: Van Nostrand Reinhold.

Getz, Donald, Darrin Joncas, and Michael Kelly 1994. Tourist shopping villages in the Calgary region. *Journal of Tourism Studies* 52–15.

Glick, Daniel. 1995. Taking the Utah desert by storm. *EcoTravel,* July/August, 12–15.

Graef, Alan R. 1977. Elements of motivation and satisfaction in the float trip experience in the Big Bend National Park. Master's thesis, Texas A&M University.

Gunn, Clare A. 1988. *Vacationscape: Designing tourist regions*, 2nd ed. New York: Van Nostrand Reinhold.

Gunn, Clare A. 1994. *Tourism planning*, 3d ed. Washington, DC: Taylor & Francis.

Gunn, Clare A. 1995. Cultural tourism planning. Paper presented at Los Caminos del Rio International Conference on the Heritage of the Lower Rio Grande, 27 January, McAllen, Texas.

HLA Consultants and ARA Consulting Group. 1994. *Ecotourism—Nature/adventure/culture: Alberta and British Columbia market demand assessment.* Location: Canadian Heritage, British Columbia, and Alberta Economic Development and Tourism.

Hollinshead, Keith. 1993. Encounters in tourism. In *VNR's encyclopedia of hospitality and tourism,* eds. Khan et al., p. 636–651. New York: Van Nostrand Reinhold.

Littrell, Mary Ann, Suzanne Baizerman, Rita Kean, Sherry Gahring, Shirley Niemeyer, Rae Kelly, and JaneAnn Stout. 1994. Souvenirs and tourism styles. *Journal of Travel Research* 33(1): 3–11.

Lundberg, Donald E. 1971. Why tourists travel. *Cornell HRA Quarterly*, February, 75–81.

Meis, Scott, and Jocelyn Lapierre. 1995. Measuring tourism's economic importance—A Canadian case study. *EIU Travel & Tourism Analyst* 2: 78–91.

Morrison, Alastair M., Philip L. Pearce, Gianna Moscardo, Nandini Nadkarni, and Joseph O'Leary. 1996. Specialist accommodation: Definition, markets served, and roles in tourism development. *Journal of Travel Research,* 35(1): 18–26.

Moskin, Bill, and Sandy Guettler. 1995. *Exploring America through its culture.* Washington, DC: President's Committee on the Arts and Humanities.

Pearce, Douglas G. 1989. *Tourism development.* New York: Longman Scientific & Technical.

Pearce, Douglas G., and Paula M. Wilson. 1995. Wildlife viewing tourists in New Zealand. *Journal of Travel Research* 34(2):19–26.

Reingold, L. 1993. Identifying the elusive tourist. *Going green,* supplement to *Tour and Travel News,* 25 October, 36–39.

Ritchie, J. R. Brent, and Donald E. Hawkins, eds. 1993. *World travel and tourism review.* Vol. 3. Wallingford, Oxon: CAB International.

Shoemaker, Stowe. 1994. Segmenting the US travel market according to benefits realized. *Journal of Travel Research,* 32(3): 8–21.

Squire, Shelagh L. 1994. The cultural values of literary tourism. *Annals of Tourism Research* 21: 103–120.

Survey Research Center. 1983. *Public participation in the arts: Final report on the 1982 survey.* Washington, DC: Survey Research Center, University of Maryland, for the National Endowment for the Arts.

Taiwan Tourism Bureau. 1995. *Monthly report on tourism, Republic of China: December.* Taiwan: Taiwan Tourism Bureau.

Task Force on Texas Nature Tourism. 1995. *Nature tourism in the Lone Star State.* Austin: Texas Parks & Wildlife Department and Texas Tourism Division.

Taylor, Gordon D. 1980. How to match plant with demand: A matrix for marketing. *Tourism Management* 1(1): 56–60.

Tighe, Anthony J. 1990. Cultural tourism in 1989. Paper presented at the 4th Annual Travel Review Conference, 5 February, Washington, DC.

Wight, Pamela. 1995. Ecotourism markets: Where are they and what are they seeking? Paper presented at Sharing Tomorrow: Exploring Responsible Tourism Conference, 24–27 September, Riding Mountain National Park, Manitoba, Canada.

Wight, Pamela. 1996. Going for the green: Marketing to ecotourists and the larger green markets. Paper presented at Opportunities in Ecotourism Conference, 23 April, Renfrew, Ontario, Canada.

World Tourism Organization. 1983. *Development of leisure time and the right to holidays.* Madrid: World Tourism Organization.

World Trade Organization. 1992. *WTO recommendations on tourism statistics.* Madrid: World Tourism Organization.

CHAPTER 4

Tourism Function: Supply

Acadia Institute of Case Studies. 1995. *Entrepreneurship: Business case studies.* Wolfville, Nova Scotia: Acadia University.

Allen, John W., David G. Armstrong, and Lawrence C. Wolken. 1979. *The foundations of free enterprise.* College Station, TX: The Center for Education and Research in Free Enterprise, Texas A&M University.

Cunningham, Lawrence F. 1994. Tourism research needs in the personal transportation modes: A 1990's perspective. In *Travel, tourism, and hospitality research.* 2d ed., eds. J. R. B. Ritchie and C. R. Goeldner, 315–325. New York: John Wiley & Sons.

Friedman, Milton, and Rose Friedman. 1980. *Free to choose.* New York: Harcourt Brace Jovanovich.

Go, Frank M. 1993. The multinational firm. In *VNR's encyclopedia of hospitality and tourism,* eds. Kahn et al., 354–365. New York: Van Nostrand Reinhold.

Green, Claudia G. 1993. The cook-chill food production process. In *VNR's encyclopedia of hospitality and tourism,* eds. Kahn, Olsen, and Var, 120–128. New York: Van Nostrand Reinhold.

Gunn, Clare A. 1994. Environmental design and land use. In *Travel, tourism, and hospitality research,* 2d eds. J. R. B. Ritchie and C. R. Goeldner, 243–258. New York: John Wiley & Sons.

Hinkson, Charles E. 1964. *Traveler profiles: A study of summer travel in Alaska during 1963 and 1964.* Juneau: Alaska Department of Economic Development and Planning.

Japan National Tourist Organization. 1991. *Tourism in Japan 1991.* Tokyo: Ministry of Transport, Japan National Tourist Organization.

Lavin, Joseph F., and Dallas S. Lunceford. 1993. Franchising and the lodging industry. In *VNR's encyclopedia of hospitality and tourism,* eds. Kahn et al., 366–372. New York: Van Nostrand Reinhold.

Lennard, Suzanne H. Crowhurst, and Henry L. Lennard. 1995. *Livable cities observed.* Carmel, CA: Gondolier.

Machlis, Gary E., ed. 1986. *Interpretive views.* Washington DC: National Parks and Conservation Association.

McCool, Audrey C. 1993. Market feasibility. In *VNR's encyclopedia of hospitality and tourism,* eds. Kahn, Olsen, and Var, 15–26. New York: Van Nostrand Reinhold.

McNulty, Robert H. 1994. Vision and goals: The essentials of leadership. In *The state of the American community,* eds. R.H. McNulty and C.A. Page, 17–30. Washington DC: Partners for Livable Communities.

McNulty, Robert H., and Patricia Wafer. 1990. Transnational corporations and tourism issues. *Tourism Management,* December, 291–295.

Passini, Romedi. 1984. *Wayfinding in architecture.* New York: Van Nostrand Reinhold.

Perdue, Richard R., and Barry E. Pitegoff. 1994. Methods of accountability research for destination marketing. In *Travel, tourism, and hospitality research.* 2d ed., eds. J. R. B. Ritchie and C. R. Goeldner, 565–571. New York: John Wiley & Sons.

Scenic America. 1991. Leaflet series including *Billboard Control Around the Country; Sign Control Helps Tourism;* and, *Fact Sheet: Billboards and the Environment.* Washington DC: Scenic America.

Scottish Tourist Board. 1994. *Strategic plan.* Edinburgh: Scottish Tourist Board.

Smith, Adam. [1776] 1930. *The wealth of nations.* 5th ed., ed. Edwin Cannan. London: Methuen & Co.

Smith, Stephen L. J. 1994. The tourism product. *Annals of Tourism Research* 21: 582–595.

Snepenger, David J., Jerry D. Johnson, and Raymond Rasker. 1995. Travel-stimulated entrepreneurial migration. *Journal of Travel Research* 34(1): 40–44.

Sullivan, Chris. 1996. Management ownership as incentive. *World's Eye View* 11(1-2):5–6.

Turler, Jerome. 1951. *The traveiler.* Gainesville, FL: Scholar's Facsimiles & Reports.

Walker, Charles, and Adrian J Smith. 1995. *Privatized infrastructure: The build operate transfer approach.* London: Thomas Telford.

Weaver, Glenn D., ed. 1991. *Tourism USA.* 3d ed. Washington, DC: U.S. Travel & Tourism Administration.

Wohlmuth, Ed. 1994. Research needs of travel retailers and wholesalers. In *Travel, tourism, and hospitality research.* 2d ed, eds. J. R. B. Ritchie and C. R. Goeldner, 263–272. New York: John Wiley & Sons.

CHAPTER 5

Attractions: First Power

Berton, Pierre. 1985. Foreword to *Reviving Main Street,* ed. Deryck Holdsworth. Toronto: University of Toronto Press.

Dulles, Foster Rhea. 1964. *Americans abroad: Two centuries of European travel.* Ann Arbor: University of Michigan Press.

Gunn, Clare A. 1965. *A concept for the design of a tourism-recreation region.* Mason, MI: B J Press.

Gunn, Clare A. 1988. *Vacationscape: Designing tourist regions,* 2nd ed. New York: Van Nostrand Reinhold.

Johnson Johnson & Roy, Inc. 1973. *Marshall: A plan for preservation.* Ann Arbor, MI: JJR.

Lew, Alan A. 1994. A framework of tourist attraction research. In *Travel, tourism, and hospitality research.* 2d ed., eds. J. R. B. Ritchie & C. R. Goeldner, 291–304. New York: John Wiley & Sons.

Mathieson, Alister, and Geoffrey Wall. 1982. *Tourism: Economic, physical and social impacts.* London: Longman.

McNulty, Robert H., and Clinton A. Page, eds. 1994. *The state of the American community.* Washington, DC: Partners for Livable Communities.

Neilson, William Allan, and Charles Jarvis Hill, eds. 1942. *The complete plays and poems of William Shakespeare.* Cambridge, MA: Houghton Mifflin.

Ralston, Jeannie. 1996. Bark grinders and fly minders tell a tale of Appalachia. *Smithsonian,* February, 26(11): 44–53.

Shepard, Paul. 1967. *Man in the landscape.* New York: Alfred A. Knopf.

Whitehand, J.W.R. 1992. *The Making of the urban landscape.* Oxford, UK: Blackwell.

Yu, Kongjian. 1995. Local people and tourists: Two models of change: A case study of Red Stone National Park, China. In *Proceedings of the 32nd IFLA World Congress,* 273–277. Meeting of the International Federation of Landscape Architects. Bangkok: Thai Association of Landscape Architects.

CHAPTER 6

Destination Development

Deveau, Linda. 1995. Success through community energy. Paper presented at conference, Chart Your Course to

Excellence. Annual meeting of the Travel Industry Association of Nova Scotia, 5–7 November, Halifax.

Gunn, Clare A. 1994. *Tourism planning,* 3d ed. Washington, DC: Taylor & Francis.

Mathieson, Alister, and Geoffrey Wall. 1982. *Tourism: Economic, physical and social impacts.* London: Longman.

Tourism Nova Scotia. 1990. *Travel guide.* Halifax: Tourism Nova Scotia.

Tourism Nova Scotia. 1995. *A strategy for tourism in Nova Scotia* (Draft). Halifax: Tourism Nova Scotia.

CHAPTER 7

Spatial Patterns

American Society of Landscape Architects and U.S. Department of Housing and Urban Development. 1977. *Barrier free site design.* Washington, DC: U.S. Government Printing Office.

Buck, Roy C., and Ted Alleman. 1979. Tourist enterprise concentration and Old Order Amish survival: Explorations in productive coexistence. *Journal of Travel Research* 18(1) 15–20.

Canada National and Historic Parks Branch, Indian Affairs and Northern Development. 1969. *National parks policy.* Ottawa: Queen's Printer.

Downie, B. K. 1984. Reflections on the National Park zoning system. *The Operational Geographer* 3: 15.

Forster, Richard R. 1973. *Planning for men and nature in the national parks.* Morges, Switzerland: International Union for Conservation of Nature and Natural Resources.

Gall, Lawrence D. 1991. Application to Minute Man of the economic impact analysis process. Letter to National Park Service regional director, North Atlantic Region, 8 July.

Gunn, Clare A. 1972. Concentrated dispersal, dispersed concentration—A pattern for saving scarce coastlines. *Landscape Architecture* 62: 133.

Gunn, Clare A. 1979. Resources management for visitors. *Fish and Wildlife News,* October–November, 20.

Gunn, Clare A., John W. Hanna, Arthur J. Perenzin, and Fred M. Blumberg. 1974. *Development of criteria for evaluating urban river settings for tourism-recreation use.* College Station: Texas Agricultural Experiment Station and Texas Water Resources Institute, Texas A&M University.

Gunn, Clare A., David J. Reed, and Robert E. Couch. 1972. *Cultural benefits from metropolitan river recreation—San Antonio prototype.* College Station: Texas Agricultural Experiment Station and Texas Water Resources Institute, Texas A&M University.

Hornback, Kenneth E. 1991. Economic value of the parks: Origin and application of the Money Genera-

tion Model, MGM. Paper presented at National Park Service North Atlantic Region Science Conference, 19 November, Newport, RI.

Sauer, Carl O. 1967. Seashore—Primitive home of man? In *Land and Life,* ed. John Leighly, 310–311. Berkeley: University of California Press.

Sax, Joseph L. 1980. *Mountains without handrails.* Ann Arbor: University of Michigan Press.

Wiedenhoeft, Ronald. 1985. Walkable cities: New approaches to environmental management. Paper presented at the Sixth Annual Pedestrian Conference, 19–20 September, Boulder, CO.

Wisconsin Rustic Roads Board. n.d. *Wisconsin rustic roads.* Madison: Wisconsin Department of Transportation.

CHAPTER 8

Techniques, Processes, and Guides

Adams, Carol. 1995. Holistic site design. Paper presented at seminar, Ecolodge Planning and Sustainable Design, 10–12 July, St. John, U.S. Virgin Islands.

Alberta Tourism. 1988. *Community tourism action plan manual.* Edmonton: Alberta Tourism.

Alberta Tourism. 1992. *Regional tourism action plan manual.* Edmonton: Alberta Tourism.

Arizona Department of Transportation. 1993. *Application procedures for designation of parkways, historic, and scenic roads in Arizona.* Phoenix: Arizona Department of Transportation.

Blank, Uel. 1989. *The community tourism industry imperative: The necessity, the opportunities, its potential.* State College, PA: Venture.

Buttner, Lisa. 1995. Renewable energy for ecolodge design. Paper presented at seminar, Ecolodge Planning and Sustainable Design, 10–23 July, St. John, U.S. Virgin Islands.

Dankittikul, Chaiyasit. 1993. The influence of the spirit world on landscape design and planning in Thailand. Dissertation, Department of Landscape Architecture, Texas A&M University, College Station, TX.

Edwards, Cheryl. 1991. Tapping into tourism. *Municipal Counsellor* 36(5): 16–17.

Fagence, Michael. 1993. "Genius loci": A catalyst for planning strategies for small rural communities. In *Community-based approaches to rural development,* eds. Bruce and Whitla, 45–68. Sackville, New Brunswick: Rural and Small Town Research and Studies Programme, Mount Allison University.

Gramann, James H., and Gail A. Vander Stoep. 1987. The effect of verbal appeals and incentives on depreciative behavior among youthful park visitors. *Journal of Leisure Research* 19(2): 69–83.

Gunn, Clare A. 1994. *An assessment of tourism potential in Newfoundland and Labrador with notes and recommendations for a planned approach to tourism development.* St. John's, Newfoundland: Hospitality Newfoundland and Labrador and Canadian Heritage.

Hill, Stephen. 1996. Letter to author, 23 January.

Lip, Evelyn. 1986. *Feng shui.* Singapore: Times Books International.

McNulty, Robert H., and Clinton A. Page, eds. 1994. *The state of the American community.* Washington, DC: Partners for Livable Communities.

Potts, Thomas D., and Allan P. C. Marsinko. 1996. *Developing naturally: An exploratory process for nature-based community tourism.* Clemson, SC: The Strom Thurmond Institute and Extension, Clemson University.

Rogers, Ala. 1993. Innovation in rural communities: A perspective from the English voluntary sector. In *Community-based approaches to rural development,* eds. Bruce and Whitla, 131–146. Sackville, New Brunswick: Rural and Small Town Research and Studies Programme, Mount Allison University.

Simonson, Lawrence R., Barbara Koth, and Glenn M. Kreag. 1988. *So your community wants travel/ tourism?* St. Paul: Minnesota Extension Service.

Sontag, William H. 1986. Skating on thin ice. In *Interpretive views,* ed. Gary E. Machlis, 77–82. Washington, DC: U.S. National Parks and Conservation Association.

Taylor, Gordon D. 1994. *How-to research manuals.* Canadian Chapter, Travel and Tourism Research Association.

Too, Lilian. 1993. *Feng shui.* Kuala Lumpur: Konsep Books.

Ulrich, Roger S. 1974. *Scenery and the shopping trip: The roadside environment as a factor in route choice.* Michigan Geographical Publication no. 12. Ann Arbor: Department of Geography, University of Michigan.

Ulrich, Roger S. 1993. Biophilia, biophobia, and natural landscapes. In *The biophilia hypothesis,* eds. Kellert and Wilson, 73–137. Washington, DC: Island Press/ Shearwater Books.

Ulrich, Roger S., and David L. Addoms. 1981. Psychological and recreational benefits of a residential park. *Journal of Leisure Research* 13(1): 43–65.

U.S. National Park Service. 1993. *Guiding principles of sustainable design.* Denver, CO: U.S. National Park Service.

U. S. National Park Service. 1995. *Renew the parks— Renewable energy in the National Park Service.* Denver, CO: National Park Service.

Weaver, Glen. 1991. *Tourism USA.* Washington, DC: U.S. Travel and Tourism Administration.

World Bank. 1991. *Environmental assessment sourcebook.* Vol. II, Sectional guidelines, environment department. Washington, DC: The World Bank.

CHAPTER 9

Conclusions and Principles

Akin, Omer, and Eleanor F. Weinel, eds. 1982. *Representation and architecture.* Silver Spring, MD: Information Dynamics.

Chimacoff, Ala. 1982. Figure, system and memory: The process of design. In *Representation and architecture,* eds. O. Akin and E. F. Weinel. Silver Spring, MD: Information Dynamics.

Ekbo, Garrett. 1969. The landscape of tourism. *Landscape* 18(2): 31.

Simonds, John Ormsbe. 1985. *Landscape architecture.* 2d ed. New York: McGraw-Hill.

Wilson, Forrest. 1984. *A graphic survey of perception and behavior for the design professions.* New York: Van Nostrand Reinhold.

Zucker, Wolfgang. 1983. The image and imagination of the architect. In *Via 6, Architecture and visual perception,* eds. Read and Doo. Cambridge, MA: MIT Press.

CHAPTER 10

Gallery of Examples

Alberta Tourism. 1994. *Environmentally sensitive facilities: Remote tourism case studies.* Edmonton: Alberta Tourism, Development Services Branch.

Andereck, Kathleen L. 1995. Arizona's rural tourism development program. *Women in Natural Resources* 16(4): 8–12.

Arizona Council for Enhancing Recreation and Tourism. 1995. *The Globe/Miami resource team report.* Phoenix: Arizona Council for Enhancing Recreation and Tourism.

Beard, Ronald E. 1993. Rural community management and conflict resolution. In *Community-based approaches to rural development,* eds. David Bruce and Margaret Whitla, 95–113. Sackville, New Brunswick: Rural and Small Town Research Studies Programme, Mount Allison University.

Conservation International. n.d. *The canopy walkway.* Washington, DC: Conservation International.

Darrow, Kit. 1995. A partnership model for nature tourism in the eastern Caribbean Islands. *Journal of Travel Research* 33(3): 48–51.

Getz, Donald. 1993. Tourism shopping villages. *Tourism Management,* February, 15–26.

Gunn, Clare A. 1979. *Tourism planning,* 1st ed. New York: Crane Russak.

Hamed, Safei El-Deen. 1991. Integrating elements of natural and cultural resources into the development of tourism in South Sinai and the west coast of the Red Sea: A strategic management plan. In *Proceedings of* the meeting of the Council of Educators in Landscape Architecture: Selected works, Ch. 9, 163–168. Washington, DC: Landscape Architecture Foundation.

Island Preservation Partnership. 1995. *Dewees Island, South Carolina—1995, the year in review.* Dewees Island, South Carolina: Island Preservation Partnership.

Kalyna Country Ecomuseum. n.d. *Kalyna Country Ecomuseum.* Lamont, Alberta: Kalyna Country Ecomuseum.

Lankford, Samuel V., and Jill Knowles Lankford. 1995. Community planning issues and tourism development. In *Proceedings of the 32nd IFLA World Congress,* 148–156. Meeting of the International Federation of Landscape Architects. Bangkok: Thai Association of Landscape Architects.

Murphy, Peter. 1985. *Tourism: A community approach.* New York: Methuen.

Shearwater Development Corporation. 1995. *Fisherman's Cove waterfront development plan.* Shearwater, Nova Scotia: Shearwater Development Corporation.

Town of Wolfville. 1994. *Destination Wolfville report.* Wolfville, Nova Scotia: Town of Wolfville.

Tracy, William A. 1994. Kalyna Country Museum. *Alberta Museums Review* 20(2): 25–27.

U.S. Department of Agriculture and National Endowment for the Arts. 1995. *Loess Hills Scenic Byway.* Washington, DC: Natural Resources Conservation Service, U.S. Department of Agriculture, and National Endowment for the Arts.

Wight, Pamela A. 1995. Ultimate adventures: An Alberta ecotourism success story. Paper presented at conference, Ecotourism in Ontario—New Business Opportunities, 24–25 November, at Sir Sandford Fleming College, Haliburton, Ontario.

A Code of Ethics for Tourists

- Travel in a spirit of humility and with a genuine desire to learn more about the people of your host country.
- Be sensitively aware of the feelings of other people, thus preventing what might be offensive behavior on your part. This applies very much to photography.
- Cultivate the habit of listening and observing, rather than merely hearing and seeing.
- Realize that often the people in the country you visit have time concepts and thought patterns different from your own; this does not make them inferior, only different.
- Instead of looking for that "beach paradise," discover the enrichment of seeing a different way of life through other eyes.

*From Ron O'Grady. 1975. Tourism, the Asian Dilemma. Paper presented at Christian Conference of Asia, as cited in David L. Edgell, 1990, *International Tourism Policy,* New York: Van Nostrand Reinhold, p. 75.

- Acquaint yourself with local customs—people will be happy to help you.
- Instead of the Western practice of knowing all the answers, cultivate the habit of listening.
- Remember that you are only one of the thousands of tourists visiting this country and so do not expect special privileges.
- If you really want your experience to be "a home away from home," it is foolish to waste money on traveling.
- When you are shopping, remember that the "bargain" you obtained was only possible because of the low wages paid to the maker.
- Do not make promises to people in your home country unless you are certain you can carry them through.
- Spend time reflecting on your daily experiences in an attempt to deepen your understanding. It has been said that what enriches you may rob and violate others.

Code of Ethics and Guidelines for Sustainable Tourism

GUIDELINES FOR THE INDUSTRY

1. Bring economic objectives into harmony with conservation of resources and environmental, social, cultural, and aesthetic values in the formulation of vision statements, mission statements, policies, plans, and the decision-making process.
2. Provide tourists with a high quality experience of our natural and cultural heritage. Facilitate as possible, meaningful contact between hosts and guests and respond to the special travel needs of diverse population segments including youth, mature citizens, and the disabled.
3. Offer tourism products and services that are consistent with community values and the surrounding environment. Reinforce and enhance landscape character, sense of place, community identity, and benefits flowing to the community as a result of tourism.
4. Design, develop, and market tourism products, facilities, and infrastructure in a manner which balances economic objectives with the maintenance and enhancement of ecological systems, cultural resources, and aesthetic resources. Achieve tourism development and marketing within a context of integrated planning.
5. Protect and enhance our natural, historic, cultural, and aesthetic resources as a legacy for present and future generations. Encourage the establishment of parks, wilderness reserves, and protected areas.
6. Practice and encourage the conservation and efficient use of natural resources including energy and water.
7. Practice and encourage environmentally sound waste and materials management including reduction, reuse, and recycling. Minimize and strive to eliminate release of any pollutant which causes environmental damage to air, water, land, flora, or wildlife.
8. Reinforce environmental and cultural awareness through marketing initiatives.
9. Encourage tourism research and education which gives emphasis to ethics, heritage preservation, and the host community; and the necessary knowledge base to ensure the economic, social, cultural, and environmental sustainability of tourism.
10. Foster greater public awareness of the economic, social, cultural, and environmental significance of tourism.
11. Act with a spirit of cooperation within the industry and related sectors to protect and enhance the environment, conserve resources, achieve balanced development, and improve the quality of life in host communities.
12. Embrace the concept of "one World" and collaborate with other nations and international bodies in the development of a socially, environmentally, and economically responsible tourism industry.

GUIDELINES FOR TOURISM INDUSTRY ASSOCIATIONS

Policy, Planning, and Decision-Making

1. Commit to excellence by incorporating sustainable tourism principles in all aspects of policy, planning, and decision-making. Develop a sustainable tourism policy statement and action plan as a demonstration of leadership. Encourage and assist members to develop policy statements and action plans with a commitment to socially, culturally, and environmentally responsible operating principles.
2. Support a proactive approach in building tourism into sustainable development strategies and ensuring that tourism resources and values are fully identified and provided for in planning and allocation processes.

*Prepared by the Tourism Industry Association of Canada and National Roundtable on the Environment of the Economy, adapted and edited by Taylor & Francis Publishers.

3. Work with respective national and provincial/territorial roundtables on the environment and the economy towards achieving sustainable development objectives for all sectors of Canada's economy.
4. Establish an evaluation and monitoring program to measure progress towards policy and action plans.

The Tourism Experience

1. Encourage our members to provide a high quality tourism experience which brings satisfaction and enrichment to visitors, hosts, and employees; greater respect, understanding and appreciation for natural and cultural resources; and which promotes an understanding and appreciation of host communities.

The Host Community

1. Encourage members to be actively involved in social, cultural, and environmental projects and events of local civic organizations and community groups.
2. Encourage the preparation of community tourism plans which incorporate sustainable tourism concepts.

Development

1. Encourage the development of tourism projects, facilities, and infrastructure which are economically, environmentally, and socially sustainable, and which secure the future of rural and urban communities.
2. Encourage efficient processes which ensure that adequate environmental information is available and considered in planning and decisions related to proposed development.
3. Encourage community involvement and informed public participation in planning tourism projects.

Natural, Cultural, and Historic Resources

1. Encourage government efforts in establishing parks, wilderness reserves, and protected areas. Collaborate with government in developing appropriate policies in relation to tourism in these areas.
2. Work with government and industry members to encourage nonconsumptive wilderness experiences, a respect for wildlife, and practices which ensure the sustainability of wildlife.
3. Encourage the development of authentic cultural tourism and heritage tourism products. Encourage recognition and respect for the values and wishes of the people whose culture and history form part of the tourism experience.
4. Encourage the preservation, restoration, and creative use of historic resources, including buildings, where economically viable.

5. Encourage the support of wilderness, cultural, and heritage programs and organizations.

Conservation of Natural Resources

1. Practice and encourage energy conservation including the purchase of energy efficient modes of transportation. Practice and encourage water conservation, the reduction of paper use, and the purchase of recycled and unbleached paper products.
2. Encourage members to select suppliers that support and practice sustainable development principles and to work with these suppliers to reduce packaging; reduce the purchase of disposable products; identify re-usable products, and products made with recyclable materials.
3. Collaborate with, and encourage other industry sectors in promoting sustainable use of air, land, water, forest, and wildlife resources.
4. Work with members to develop a collective tourism position on significant natural resource issues.

Environmental Protection

1. Practice and encourage environmentally sound waste and materials management including reduction, recycling, and re-use. Encourage efforts to reduce pollution in all forms.
2. Encourage members to work with suppliers to develop environmentally friendly products, to reduce packaging, and to develop re-usable shipping containers.

Marketing

1. Encourage the development and promotion of tourism products and activities that enhance environmental and cultural awareness.
2. Encourage promotion which highlights Canada's natural, cultural, and historic resources.

Research and Education

1. Encourage government and industry research that advances knowledge and practices related to sustainable tourism including the economic, social, cultural, and environmental impacts of proposed projects. Support research for the development of sustainable tourism indicators. Encourage research which monitors consumer response to sustainable tourism initiatives.
2. Encourage the development and introduction of sustainable tourism concepts, principles, and practices in tourism and education programs at all levels.
3. Host and encourage educational seminars related to sustainable tourism and relevant aspects of environmental, cultural, and historic resources for management and employees within the industry.

Public Awareness

1. Support the development of public policy, and industry, government, and educational initiatives which increase environmental and cultural awareness, the concept of sustainable development, and the potential contribution of tourism towards this end.
2. Provide members with knowledge, and promote awareness regarding the Code of Ethics for Tourists, the Code of Ethics for the Industry, and Recommended Guidelines for respective sectors of the industry.
3. Identify and publicize success stories among members.

Industry Cooperation

1. Cooperate, and encourage members to cooperate, within the industry, with government, and with other organizations working towards sustainable tourism and improved quality of life in host communities and tourism regions.
2. Encourage participation in events such as Tourism Awareness Week, World Tourism Day, Parks Day, Heritage Day, Environment Week, and the UNESCO Decade for Cultural Development.

GUIDELINES FOR ACCOMMODATION

Policy, Planning, and Decision-Making

1. Commit to excellence in incorporating sustainable principles in all aspects of policy, planning, and decision-making.
2. Prepare an environmental policy statement and action plan. Establish an environmental committee to develop programs and generate staff support.
3. Establish a monitoring and evaluation program to assess progress.

Guests

1. Reinforce environmental awareness among guests. Inform guests of environmental programs and advise how they can cooperate in conserving energy and water, and the recycling of waste.

The Host Community

1. Encourage the development of community and regional infrastructure for the collection, storage, and processing of recyclable materials.
2. Donate excess food to local shelters and community groups as possible within the framework of applicable health regulations.

3. Support cultural and environmental programs of community groups and organizations.

Development

1. Respect natural and cultural surroundings in the scale, siting, design, and operation of new facilities, expansions and renovations. As possible, enhance the quality of the landscape.
2. Design and develop facilities taking into consideration efficient use of energy and materials, the sustainable use of natural resources, the minimization of adverse environmental impacts and waste generation, and the safe and responsible disposal of residual wastes.
3. Consult with the Royal Canadian Institute of Architecture to select materials that are non-toxic and which are least harmful in their harvesting, mining, or manufacture, use and disposal.

Natural, Cultural, and Historic Resources

1. Consider the use of local culture and local artists to enhance new and renovated buildings and the use of local materials in construction.
2. Commit to the preservation and restoration of historic hotels when economically viable.

Conservation of Natural Resources

1. Practice energy conservation in all areas including heating, air conditioning, and lighting. Consider the use of solar energy in new facilities and as possible, upgrade to more energy efficient systems in all facilities.
2. Encourage the use of public transportation and ride-sharing among employees and facilitate walking, jogging, bicycling to work by installing showers, lockers, and bicycle racks.
3. Practice water conservation and install, as possible, water conserving fixtures.
4. Purchase recycled and unbleached paper products for guest rooms, dining facilities, and office use.
5. Conduct periodic energy and water audits.

Environmental Protection

1. Minimize and try to eliminate release of any pollutants.
2. Minimize the generation of waste through reduction, re-use, and recycling. Dispose of waste in an environmentally safe manner.
3. Seek out practical options for the diversion of waste to useful purposes such as composting and conservation of food leftovers to animal feed.
4. Use environmentally friendly guest amenities.
5. Purchase supplies in bulk as practical and utilize dis-

pensers in dining facilities for condiments, soft drinks, and dairy products where health regulations allow.

6. Work with suppliers to develop environmentally friendly products, to reduce packaging, and to develop re-usable shipping containers.
7. Periodically check air conditioning systems and refrigeration units for leaks to prevent loss of freon.
8. Reduce indoor air pollution by installing air cleaning plants.

Marketing

1. Reflect environmental initiatives in marketing and promotion programs.

Research and Education

1. Support and encourage research related to the reduction, recycling, and re-use of wastes and the advancement of knowledge contributing to sustainable tourism including the development of sustainable tourism indicators. Support and initiate research which monitors consumer response to sustainable tourism initiatives.
2. Educate and motivate staff regarding the implementation of environmental policies with the aim of instilling an environmentally and culturally caring work ethic consistent with health and safety factors.
3. Remain informed of significant developments regarding relevant environmental practices.

Public Awareness

1. Encourage the development of public policy, and industry, government and educational initiatives which increase environmental and cultural awareness, understanding of the concept of sustainable development and the contribution of tourism towards these ends.

Industry Cooperation

1. Cooperate within the industry, with government, and with other organizations working towards the aim of sustainable tourism and an improved quality of life in destination areas.
2. Cooperate with community organizations and other local industries in achieving sustainable development goals of the community.

Global Village

1. Encourage participation in events such as Tourism Awareness Week, World Tourism Day, Heritage Day, Parks Day, Earth Day, Environment Week, and UNESCO World Decade for Cultural Development.

GUIDELINES FOR FOOD SERVICE

Policy, Planning, and Decision-Making

1. Develop environmental policies and consider environmental factors in all aspects of planning and decision-making.
2. Identify and comply with environmental laws and regulations as they relate to food and beverage.
3. Conduct periodic environmental (including waste) audits and evaluations to ensure compliance with policy and regulations.

The Consumer

1. Inform customers of steps being taken to protect the environment and to minimize waste.
2. Invite the cooperation of customers in achieving environmental aims and provide guidelines as appropriate.

The Community

1. Encourage the development of community and regional infrastructure for the collection, storage, and processing of recyclable materials.

Development

1. Consider natural and cultural surroundings in the design, siting, and operation of new facilities, expansions or renovations. Respect and as possible, enhance the quality of the landscape/streetscape.
2. Design and develop facilities taking into consideration efficient use of energy and materials; the sustainable use of natural resources; the minimization of adverse environmental impacts and waste generation; and the safe and responsible disposal or diversion of residual wastes.

Natural, Cultural, and Historic Resources

1. Encourage the preservation, restoration, and creative use of historic buildings where economically feasible.

Conservation of Natural Resources

1. Practice energy conservation and utilize energy efficient equipment for food storage and preparation. Keep heating and air conditioning equipment well-maintained and upgrade systems as possible for increased efficiency.
2. Seek to conserve water and to reduce the use of paper. Purchase recycled and unbleached paper products such as towels and napkins as possible.

Environmental Protection

1. Minimize the generation of all waste through reduction, re-use and recycling. Seek out practical options for the diversion of waste to useful purposes such as composting and conversion to animal feed.
2. Purchase supplies in bulk as practical and utilize dispensers for condiments, soft drinks, and other items where health regulations allow.
3. Work with suppliers to reduce packaging and to develop re-usable shipping containers.
4. Periodically check refrigeration units and air conditioners for leaks to prevent loss of freon.

Marketing

1. Reflect environmental initiatives in marketing and promotion programs.

Research and Education

1. Support and encourage research related to the reduction, recycling, and re-use of all wastes including organics and packaging, and diversion alternatives such as composting and conversion to animal feed.
2. Educate and involve employees regarding the content and purpose of environmental policies. Encourage a positive team effort in support of environmental policies and practices consistent with health and safety.

Public Awareness

1. Encourage industry efforts to promote an understanding of environmental issues related to food and beverage; industry efforts to deal with them in a positive way; and how consumers can cooperate in support of these initiatives.

Industry Cooperation

1. Collaborate through appropriate industry associations in encouraging governments to establish facilities and infrastructure for the recycling of all food and beverage related wastes including grease, organics, paper, cardboard, metals, plastics, and glass.
2. Support industry efforts to collaborate with government in the development of appropriate environmental and waste management policies. Cooperate in the implementation of voluntary guidelines established by the National Packaging Protocol.

Global Village

1. Participate in events such as Parks Day, Earth Day, and Environment Week.

GUIDELINES FOR TOUR OPERATORS

Policy, Planning, and Decision-Making

1. Prepare a sustainable tourism policy statement with a commitment to socially, culturally, and environmentally responsible practices in planning and conducting tours.
2. Conduct periodic monitoring and evaluation of tours to ensure compliance with the sustainable tourism policy statement and regulations of the host destination.

The Tourism Experience

1. Provide a high quality tourism experience which brings satisfaction and enrichment to clients, brings greater knowledge and appreciation of natural and cultural heritage, and promotes an understanding and appreciation of host communities.
2. Prepare clients with pre-trip information regarding the host destination including local customs, traditions, and proper etiquette. On tours to foreign lands, provide tour participants with phrases for basic communication in the local language.
3. Facilitate, as possible and appropriate, meaningful contact between hosts and guests.

The Host Community

1. Offer tours that are consistent with host community values; reinforce community identity; and provide commensurate benefits to the host community.
2. As possible, gain an understanding of the visions and plans of host destinations, and the context of tourism within community goals and aspirations. Seek local perspectives in planning interpretive programs. Where possible and appropriate, hire local guides and operators, and support local businesses and service providers.
3. Encourage and support host community environmental and cultural initiatives and efforts towards sustainable tourism.

Development

1. Develop tour products which provide authentic experiences while respecting the values and wishes of all people whose culture and history form part of the tourism experience.
2. Support tourism development which is compatible with the culture, values, and lifestyles of host communities and sensitive to environmental surroundings.
3. Encourage and support, where appropriate, local entrepreneurial tourism development.

Natural, Cultural, and Historic Resources

1. Cultivate a fuller understanding and appreciation among tour participants for the people, their culture,

traditions, and history and the natural surroundings of host destinations.
2. Foster a respect for nature and wildlife. Have knowledge of, and respect local guidelines and regulations in parks and wilderness areas.
3. Encourage and support local organizations seeking to preserve and enhance natural, cultural, and historic resources.

Conservation of Natural Resources

1. Practice, and encourage tour participants to practice the conservation of resources including energy and water. Select as possible, energy efficient modes of transportation.
2. As possible, select service providers who practice conservation of resources and a respect for the environment.
3. Use recycled and unbleached paper products for promotional literature.

Environmental Protection

1. Cultivate among tour participants an understanding of their role in preventing pollution in all forms; the proper disposal of waste; and the protection of wilderness, cultural, archeological, and historic resources.
2. Discourage the purchase of products or services which threaten wildlife and plant populations.

Marketing

1. Develop and market products that encourage visitors to wisely enjoy natural, cultural, and historic resources; and to learn about the culture, lifestyle, and traditions of the people they visit.

2. Provide reliable information in promotional material. As possible, reinforce environmental and cultural awareness in marketing programs.

Research and Education

1. Encourage and participate in research which contributes to the aim of sustainable tourism including the development of sustainable tourism indicators.
2. Encourage the education of tour guides which includes appropriate emphasis on social, cultural, environmental, and historic features of host destinations.

Public Awareness

1. Provide clients with a copy of "Code of Ethics for Tourists."
2. Support industry efforts to foster greater awareness of the economic, social, cultural, and environmental significance of tourism.

Industry Cooperation

1. Cooperate within the industry and with host destinations in achieving a high standard of sustainable tourism and an improved quality of life in destination areas.

Global Village

1. Support participation in events such as Tourism Awareness Week, World Tourism Day, Heritage Day, Parks Day, Environment Week, and the UNESCO Decade for Cultural Development.
2. Provide awareness of the role of tourism in promoting international understanding and cooperation.

Index

Vacation home sites, 76
Valley of the Kings, 6
Vancouver, British Columbia, 16
Vander Stoep, Gail A., 91
Venice, 5
Viewshed, 96
Visas, 22
Visiting friends, relatives, 72
Visitor (*see also* Traveler)
  characteristics, 21
  experience, 19, 30, 57
  satisfaction, 6, 51, 54
Voortrekker Monument, 7

Wafer, Patricia, 36
Waikiki Beach, 87
Walker, Charles, 41
Walkway, 35
Wall, Geoffrey, 47, 67
Walt Disney World, 6, 22
Washington, D.C., 43
Waste management, 4, 10, 91
Water tours, 71
Water supply, 6
Wayfinding, 35

Wealth of Nations, 39
Weaver, Glenn, 6, 40, 97
Weinel, Eleanor F., 106
Werkmeister, Hans Friedrich, 6
Western Australia Tourism Commission,
  17
Whitehand, J. W. R., 51
Wiedenhoeft, Ronald, 72
Wight, Pamela, 4, 24, 117
Wilderness Society (U.S.), 24
Wildlife, 7
Williamsburg, Virginia, 40, 51
Wilson, Forrest, 107
Wind (electrical) systems, 91
Wisconsin Rustic Roads Board, 69
Wohlmuth, Ed., 36
Wolken, Lawrence, 39
Workshop, 92
World Bank, 18, 93, 95
World Fair (Venice) 5
World Park, China, 5
World Tourism Organization (WTO), 13,
  21, 22, 24
World trade, 1
World Travel & Tourism Council
  (WTTC), 1, 8, 9, 13

World Travel and Tourism Environmental
  Research Centre (WTTERC), 9,
  18
Wright, Michael, 14

Xiang, Sun Xiao, 5

Yarmouth County Tourist Association,
  65
Yarmouth Development Corporation,
  65
Yellowstone National Park, 17, 20, 40
Yin, Yang, 89
Young Men's/Women's Christian
  Association, 75
Yu, Kongjian, 57

Zoning, 4
  by time, 35
  legal foundation, 90
  roadside, 37
  statutes, laws, 4, 69
Zucker, Wolfgang, 112